Hadoop实战手册

[美] Jonathan R. Owens　　Jon Lentz　　Brian Femiano　著
傅杰　赵磊　卢学裕　译

人民邮电出版社
北京

图书在版编目（CIP）数据

Hadoop实战手册 /（美）欧文斯（Owens, J.R.），（美）伦茨（Lentz, J.），（美）费米亚诺（Femiano, B.）著；傅杰，赵磊，卢学裕译. -- 北京：人民邮电出版社，2014.1（2017.8重印）
书名原文：Hadoop real-world solutions cookbook
ISBN 978-7-115-33795-5

Ⅰ. ①H… Ⅱ. ①欧… ②伦… ③费… ④傅… ⑤赵… ⑥卢… Ⅲ. ①数据处理软件—技术手册 Ⅳ. ①TP274-62

中国版本图书馆CIP数据核字（2013）第277092号

版权声明

Copyright ©2013 Packt Publishing. First published in the English language under the title *Hadoop Real-World Solutions Book*.

All Rights Reserved.

本书由英国Packt Publishing公司授权人民邮电出版社出版。未经出版者书面许可，对本书的任何部分不得以任何方式或任何手段复制和传播。

版权所有，侵权必究。

◆ 著　　[美] Jonathan R. Owens　Jon Lentz　Brian Femiano
　 译　　傅 杰　赵 磊　卢学裕
　 责任编辑　杨海玲
　 责任印制　程彦红

◆ 人民邮电出版社出版发行　北京市丰台区成寿寺路11号
　 邮编　100164　电子邮件　315@ptpress.com.cn
　 网址　http://www.ptpress.com.cn
　 固安县铭成印刷有限公司印刷

◆ 开本：800×1000　1/16
　 印张：16.25
　 字数：242千字　　　　　　　2014年1月第1版
　 印数：3 501—3 700册　　　2017年8月河北第2次印刷

著作权合同登记号　图字：01-2013-4468号

定价：59.00元
读者服务热线：(010)81055410　印装质量热线：(010)81055316
反盗版热线：(010)81055315

内容提要

这是一本 Hadoop 实用手册，主要针对实际问题给出相应的解决方案。本书特色是以实践结合理论分析，手把手教读者如何操作，并且对每个操作都做详细的解释，对一些重要的知识点也做了必要的拓展。全书共包括 3 个部分，第一部分为基础篇，主要介绍 Hadoop 数据导入导出、HDFS 的概述、Pig 与 Hive 的使用、ETL 和简单的数据处理，还介绍了 MapReduce 的调试方式；第二部分为数据分析高级篇，主要介绍高级聚合、大数据分析等技巧；第三部分为系统管理篇，主要介绍 Hadoop 的部署的各种模式、添加新节点、退役节点、快速恢复、MapReduce 调优等。

本书适合各个层次的 Hadoop 技术人员阅读。通过阅读本书，Hadoop 初学者可以使用 Hadoop 来进行数据处理，Hadoop 工程师或者数据挖掘工程师可以解决复杂的业务分析，Hadoop 系统管理员可以更好地进行日常运维。本书也可作为一本 Hadoop 技术手册，针对要解决的相关问题，在工作中随时查阅。

译者序

随着 Hadoop 技术在互联网公司的广泛应用，普及程度越来越高，自学 Hadoop 的 Java 程序员也越来越多。大多数人（包括译者本人）自学 Hadoop 的都是从"部署 Hadoop 环境+运行 WordCount 例子"开始，而且大多数自学者也都终止在 WordCount。因没有具体的应用场景而感到学习没有方向，没有成就感。

本书特色是以实践结合理论分析，手把手教读者如何操作，并且对每个操作都做详细的解释，对一些重要的知识点也做了必要的拓展。此外，书中的教学源代码都可以在官网上下载到。本书整体可分成 3 部分，第一部分为基础篇包含第 1 章、第 2 章、第 3 章、第 4 章、第 8 章内容，主要介绍 Hadoop 数据导入导出、HDFS 的概述、Pig 与 Hive 的使用、ETL 和简单的数据处理，还介绍了 MapReduce 的调试方式；第二部分为数据分析高级篇包含第 5 章、第 6 章、第 7 章、第 10 章内容，主要介绍高级聚合、大数据分析等技巧；第三部分为系统管理篇，包含第 9 章，主要介绍 Hadoop 的部署的各种模式、添加新节点、退役节点、快速恢复、MapReduce 调优等。

如果你是 Hadoop 初学者，建议你先阅读第一部分内容，完成这部分内容的学习以后，你基本上可以使用 Hadoop 来进行数据处理。

如果你已经是 Hadoop 工程师或者数据挖掘工程师，可以系统地学习第二部分内容，当然也可以根据需要进行查阅学习。完成这部分内容的学习，有助于解决一些复杂的业务分析。

如果你是 Hadoop 系统管理员，建议你阅读第三部分内容，当然你也可以阅读第一部分的内容，这样更有助于进行日常运维。

本书也可作为一本手册，在教学、工作中随时查阅，解决相关问题。

译者简介

傅杰 硕士，毕业于清华大学高性能所，现就职于优酷土豆集团，任数据平台架构师，负责集团大数据基础平台建设，支撑其他团队的存储与计算需求，包含 Hadoop 基础平台、日志采集系统、实时计算平台、消息系统、天机镜系统等。个人专注于大数据基础平台架构及安全研究，积累了丰富的平台运营经验，擅长 Hadoop 平台性能调优、JVM 调优及诊断各种 MapReduce 作业，还担任 China Hadoop Submit 2013 大会专家委员、优酷土豆大数据系列课程策划&讲师、EasyHadoop 社区讲师。

赵磊 硕士，毕业于中国科学技术大学，现就职于优酷土豆集团，任数据挖掘算法工程师，负责集团个性化推荐和无线消息推送系统的搭建和相关算法的研究。个人专注于基于大数据的推荐算法的研究与应用，积累了丰富的大数据分析与数据挖掘的实践经验，对分布式计算和海量数据处理有深刻的认识。

卢学裕 硕士，毕业于武汉大学，曾供职腾讯公司即通部门，现就职于优酷土豆集团，担任大数据技术负责人，负责优酷土豆集团大数据系统平台、大数据分析、数据挖掘和推荐系统。有丰富的 Hadoop 平台使用及优化经验，尤其擅长 MapReduce 的性能优化。基于 Hadoop 生态系统构建了优酷土豆的推荐系统，BI 分析平台。

前言

本书能帮助开发者更方便地使用 Hadoop，从而熟练地解决问题。读者会更加熟悉 Hadoop 相关的各种工具从而进行最佳的实践。

本书指导读者使用各种工具解决各种问题。这些工具包括：Apache Hive、Pig、MapReduce、Mahout、Giraph、HDFS、Accumulo、Redis 以及 Ganglia。

本书提供了深入的解释以及代码实例。每章的内容包含一组问题集的描述，并对面临的技术挑战提出了解决方案，最后完整地解决了这些问题。每节将单一问题分解成不同的步骤，这样更容易按照步骤执行相关操作。本书覆盖的内容包括：关于 HDFS 的导入、导出数据，使用 Giraph 进行图分析，使用 Hive、Pig 以及 MapReduce 进行批量数据分析，使用 Mahout 进行机器学习方法，调试并修改 MapReduce 作业的错误，使用 Apache Accumulo 对结构数据进行列存储与检索。

本书的示例中涉及的 Hadoop 技术同样也可以应用于读者自己所面对的问题。

本书涵盖哪些内容

第 1 章 "Hadoop 分布式文件系统——导入和导出数据"，包含了从一些流行的数据库导入导出数据的方法，包括 MySQL、MongoDB、Greenplum 以及 MSSQL Server。此外，还包括一些辅助工具，例如 Pig、Flume 以及 Sqoop。

第 2 章 "HDFS"，介绍从 HDFS 读入或写出数据，介绍了如何使用不同的序列化库，包含 Avro、Thrift 以及 Protocol Buffers。同样包含如何设置数据块大小、备份数以及是否需要进行 LZO 压缩。

第 3 章 "抽取和转换数据"，包含对不同数据源类型进行基本的 Hadoop ETL 操作。不同的工具包括 Hive、Pig 以及 MapRedcue JAVA API，用于批量处理数据，输出一份或多份转换数据。

第 4 章 "使用 Hive、Pig 以及 MapReduce 处理常见的任务"，关注于如何利用这些工具提供的函数快速解决不同种类的问题。其中包括字符串连接，外部表的映射，简单表的连接，自定义函数以及基于集群进行分发操作。

第 5 章 "高级连接操作"，介绍了 MapReduce、Hive 以及 Pig 中复杂而有效的连接技术。章节的内容包括 Pig 中的归并连接、复制连接以及倾斜连接。同样也包含 Hive 的 map 端连接以及全外连接的内容。同时本章也包括对于外部数据存储，如何使用 Redis 进行数据连接。

第 6 章 "大数据分析"，介绍了如何使用 Hadoop 解答关于你的不同数据的各种查询。一些关于 Hive 的例子将展示在不同的分析中如何正确地实现并重复使用用户自定义函数（UDF）。此外其中有关于 Pig 的两小节，介绍了对于 Audioscrobbler 数据集的各类分析。关于 MapReduce Java API 的一小节介绍了 Combiner 的使用。

第 7 章 "高级大数据分析"，展示了使用 Apache Giraph 以及 Mahout 处理不同类型的图分析和机器学习问题。

第 8 章 "调试"，帮助你定位并测试 MapReduce 作业。这些例子使用 MRUnit 以及更利于测试的本地模式。此外还强调了使用 counter 以及更新任务状态的重要性，这样做有助于监控 MapReduce 作业。

第 9 章 "系统管理"，主要关注如何性能调优 Hadoop 中不同的配置项。下面的内容包含在内：基本的初始化，XML 配置项调整，定位坏数据节点，处理 NameNode 故障，使用 Ganglia 进行性能监控。

第 10 章 "使用 Apache Accumulo 进行持久化"，展示了使用 NoSQL 数据存储 Apache Accumulo 带来的很多特性和功能。这些章节利用了这些独特的特性，包括 iterator、combiner、扫描授权以及约束，同时也给出了对于有效建立地理空间行健值以及使用 MapReduce 执行批量分析的例子。

阅读需要的准备

读者需要访问一个伪分布式（单台机器）或完全分布式（多台机器）的集群用于执行

本书中的例子。章节中使用的各种工具需要在集群上安装并正确地配置。此外，本书提供的代码使用不同的语言编写，因此读者能访问的机器最好安装了合适的开发工具。

读者范围

本书使用简要的代码作为例子，展示了可以使用 Hadoop 解决的各类现实中的问题。本书的目的是使不同水平的开发者都能方便地使用 Hadoop 及其工具。Hadoop 的初学者能通过本书加速学习进度并了解现实中 Hadoop 应用的例子。对于有更多经验的 Hadoop 开发者，通过本书，会对许多工具以及技术有新的认识或有一个更清晰的框架，这些东西之前可能你只听说过但并没有真正理解其中的价值。

书中的约定

在你阅读本书时，你会发现书中有各种样式的文本，这些不同样式的文本是用来区分不同类型的信息的。下面是一些不同样式文本的实例，以及相应的说明。

文本中的代码如下所示："所有的 Hadoop 文件系统 shell 命令行都使用统一的形式 `hadoop fs -COMMAND`。"

代码块如下所示：

```
weblogs = load '/data/weblogs/weblog_entries.txt' as
            (md5:chararray,
             url:chararray,
             date:chararray,
             time:chararray,
             ip:chararray);

md5_grp = group weblogs by md5 parallel 4;

store md5_grp into '/data/weblogs/weblogs_md5_groups.bcp';
```

如果我们希望让代码块中的一特殊部分引起你的注意，相关行或条目会以黑体印刷：

```
weblogs = load '/data/weblogs/weblog_entries.txt' as
            (md5:chararray,
             url:chararray,
             date:chararray,
             time:chararray,
```

```
                    ip:chararray);

md5_grp = group weblogs by md5 parallel 4;

store md5_grp into '/data/weblogs/weblogs_md5_groups.bcp';
```

命令行输入或输出都是以如下样式书写的：

```
hadoop distcp -m 10 hdfs://namenodeA/data/weblogs hdfs://namenodeB/data/weblogs
```

新的术语以及重要文字会以加粗的字体出现。你在屏幕上（如菜单或者对话框中）见到的文字都会是像后面这样出现在正文中："为了构建 JAR 文件，下载 Jython Java 安装程序，并执行该程序，从安装菜单中选择 **Standalone** 选项。"

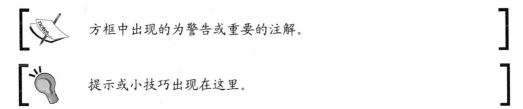

方框中出现的为警告或重要的注解。

提示或小技巧出现在这里。

读者反馈

我们总是欢迎来自读者的反馈。请告诉我们你觉得这本书怎么样，你喜欢哪些内容，不喜欢哪些内容。读者的反馈对于帮助我们写出那些对读者真正有用的内容至关重要。

如果是给我们反馈一些普通的信息，你可以给 feedback@packtpub.com 这个邮箱发一封邮件即可，记得在你邮件的标题中提及相应的书名。

如果你是某一方面的专家并且对于写作或者撰稿有兴趣的话，你可以访问 www.packtpub.com/authors，读读我们的作者指南。

客户支持

现在你成为了 Packt 出版社的一名尊敬的用户，为使你的购买物超所值，我们为你准备了一系列的东西。

下载本书中的示例代码

你可以登录账号下载到所有从 http://www.packtpub.com 购买的 Packt 的书的示例代码文件。如果你从别的地方购买了此书，可以访问 http://www.packtpub.com/support 登记信息，文件将通过邮件直接发给你。

勘误

尽管我们已经非常小心谨慎，以确保内容的准确性，错误还是不可避免。如果你在书中发现了错误（这种错误可能是文字或者代码方面的），若你能向我们报告这些错误，我们将感激不尽。这样做，可以使其他读者免受这些错误带来的困扰，帮助改善该书的下一个版本。如果你发现了什么错误，请访问 http://www.packtpub.com/support，选择书名，点击"提交错误"链接，然后输入你发现错误的详细内容，通知我们。一旦你指出的错误得到确认，你提交的内容就会被采纳，并加入一个已经存在的勘误列表中。你也可以访问这个链接 http://www.packtpub.com/support，选择书名，查看相应书本已有的勘误表。

关于盗版

所有媒体的版权材料在互联网上被盗版是一个日趋严重的问题。在 Packt，我们对于版权与许可证的保护工作是十分看重的。如果你发现任何非法复制我们作品的现象，无论以任何形式，只要在互联网上，请提供给我们对应的网络地址或网站名称，我们会立即对其行为进行纠正。

请通过 copyright@packtpub.com 联系我们，并附上可疑盗版材料的链接。

我们感谢你对保护作者权益的帮助，你的协助同时也保障了我们带给你更多有价值内容的能力。

如果你有疑问

如果你对书的某些方面有疑问，你可以通过 questions@packtpub.com 联系我们，我们会尽最大的努力为你解答。

作者简介

Jonathan R. Owens：软件工程师，拥有 Java 和 C++ 技术背景，最近主要从事 Hadoop 及相关分布式处理技术工作。

目前就职于 comScore 公司，为核心数据处理团队成员。comScore 是一家知名的从事数字测评与分析的公司，公司使用 Hadoop 及其他定制的分布式系统对数据进行聚合、分析和管理，每天处理超过 400 亿单的交易。

> 感谢我的父母 James 和 Patricia Owens，感谢他们对我的支持以及从小给我介绍科技知识。

Jon Lentz：comScore 核心数据处理团队软件工程师。comScore 是一家知名的在线受众测评与分析的公司。他更倾向于使用 Pig 脚本来解决问题。在加入 comScore 之前，他主要开发优化供应链和分配固定收益证券的软件。

> 我的女儿 Emma 出生在本书写作的过程中。感谢她深夜的陪伴！

Brian Femiano：本科毕业于计算机科学专业，并且从事相关专业软件开发工作 6 年，最近两年主要利用 Hadoop 构建高级分析与大数据存储。他拥有商业领域的相关经验，以及丰富的政府合作经验。他目前就职于弗吉尼亚的 Potomac Fusion 公司，这家公司主要从事可扩展算法的开发，并致力于学习并改进政府领域中最先进和最复杂的数据集。他通过教授课程和会议培训在公司内部普及 Hadoop 和云计算相关的技术。

我要感谢合著者的耐心，以及他们为你们可以在书上看到的代码做出的努力。同时也要感谢 Potomac Fusion 的许多同事，他们驾驭最前沿技术的才能和激情，以及促进知识传播的精神一直鼓舞着我。

审阅者简介

Edward J. Cody：作家、演讲师，大数据仓库、Oracle 商业智能和 Hyperion EPM 实现的专家，是 *The Business Analyst's Guide to Oracle Hyperion Interactive Reporting 11* 的作者，以及 *The Oracle Hyperion Interactive Reporting 11 Expert Guide* 的合著者。在过去的职业生涯中，他给商业公司和联邦政府都做过咨询，目前主要从事大型 EPM、BI 和大数据仓库的实现研究。

> 本书的作者做了很好的工作，同时我要感谢 Packt 出版社让我有机会参与本书的编辑出版工作。

Daniel Jue：Sotera Defense Solutions 公司的高级软件工程师，Apache 软件基金会成员之一，他曾在和平与战乱地区为 ACSIM、DARPA 和许多联邦机构工作，致力于揭示蕴藏在"大数据"背后的动力学与异常现象。Daniel 拥有马里兰大学帕克分校计算机专业的学士学位，并同时专注于物理学与天文学的研究，他目前的兴趣是将分布式人工智能技术应用到自适应异构的云计算中。

> 感谢我漂亮的妻子 Wendy，以及我的双胞胎儿子 Christopher 和 Jonathan，在我审阅本书的过程中，他们给予了我关爱与耐心。非常感激 Brian Femiano、Bruce Miller 和 Jonathan Larson，他们让我接触到了许多伟大的想法、观点，使我深受启发。

Bruce Miller：Sotera Defense Solutions 公司高级软件工程师，目前受雇于美国国防部高级研究计划署（DARPA），并专注于大数据的软件开发 10 余年。工作之余喜欢用 Haskell 和 Lisp 语言进行编程，并将其应用于解决一些实际的问题。

目录

第 1 章 Hadoop 分布式文件系统——导入和导出数据 ... 1
- 1.1 介绍 ... 1
- 1.2 使用 Hadoop shell 命令导入和导出数据到 HDFS ... 2
- 1.3 使用 `distcp` 实现集群间数据复制 ... 7
- 1.4 使用 Sqoop 从 MySQL 数据库导入数据到 HDFS ... 9
- 1.5 使用 Sqoop 从 HDFS 导出数据到 MySQL ... 12
- 1.6 配置 Sqoop 以支持 SQL Server ... 15
- 1.7 从 HDFS 导出数据到 MongoDB ... 17
- 1.8 从 MongoDB 导入数据到 HDFS ... 20
- 1.9 使用 Pig 从 HDFS 导出数据到 MongoDB ... 23
- 1.10 在 Greenplum 外部表中使用 HDFS ... 24
- 1.11 利用 Flume 加载数据到 HDFS 中 ... 26

第 2 章 HDFS ... 28
- 2.1 介绍 ... 28
- 2.2 读写 HDFS 数据 ... 29
- 2.3 使用 LZO 压缩数据 ... 31
- 2.4 读写序列化文件数据 ... 34
- 2.5 使用 Avro 序列化数据 ... 37
- 2.6 使用 Thrift 序列化数据 ... 41
- 2.7 使用 Protocol Buffers 序列化数据 ... 44
- 2.8 设置 HDFS 备份因子 ... 48

2.9 设置 HDFS 块大小 49

第 3 章 抽取和转换数据 51

3.1 介绍 51
3.2 使用 MapReduce 将 Apache 日志转换为 TSV 格式 52
3.3 使用 Apache Pig 过滤网络服务器日志中的爬虫访问量 54
3.4 使用 Apache Pig 根据时间戳对网络服务器日志数据排序 57
3.5 使用 Apache Pig 对网络服务器日志进行会话分析 59
3.6 通过 Python 扩展 Apache Pig 的功能 61
3.7 使用 MapReduce 及二次排序计算页面访问量 62
3.8 使用 Hive 和 Python 清洗、转换地理事件数据 67
3.9 使用 Python 和 Hadoop Streaming 执行时间序列分析 71
3.10 在 MapReduce 中利用 `MultipleOutputs` 输出多个文件 75
3.11 创建用户自定义的 Hadoop Writable 及 InputFormat 读取地理事件数据 78

第 4 章 使用 Hive、Pig 和 MapReduce 处理常见的任务 85

4.1 介绍 85
4.2 使用 Hive 将 HDFS 中的网络日志数据映射为外部表 86
4.3 使用 Hive 动态地为网络日志查询结果创建 Hive 表 87
4.4 利用 Hive 字符串 UDF 拼接网络日志数据的各个字段 89
4.5 使用 Hive 截取网络日志的 IP 字段并确定其对应的国家 92
4.6 使用 MapReduce 对新闻档案数据生成 n-gram 94
4.7 通过 MapReduce 使用分布式缓存查找新闻档案数据中包含关键词的行 98
4.8 使用 Pig 加载一个表并执行包含 `GROUP BY` 的 `SELECT` 操作 102

第 5 章 高级连接操作 104

5.1 介绍 104
5.2 使用 MapReduce 对数据进行连接 104
5.3 使用 Apache Pig 对数据进行复制连接 108
5.4 使用 Apache Pig 对有序数据进行归并连接 110
5.5 使用 Apache Pig 对倾斜数据进行倾斜连接 111
5.6 在 Apache Hive 中通过 map 端连接对地理事件进行分析 113
5.7 在 Apache Hive 通过优化的全外连接分析地理事件数据 115

5.8	使用外部键值存储（Redis）连接数据	118

第 6 章　大数据分析 ·· 123

6.1	介绍	123
6.2	使用 MapReduce 和 Combiner 统计网络日志数据集中的独立 IP 数	124
6.3	运用 Hive 日期 UDF 对地理事件数据集中的时间日期进行转换与排序	129
6.4	使用 Hive 创建基于地理事件数据的每月死亡报告	131
6.5	实现 Hive 用户自定义 UDF 用于确认地理事件数据的来源可靠性	133
6.6	使用 Hive 的 map/reduce 操作以及 Python 标记最长的无暴力发生的时间区间	136
6.7	使用 Pig 计算 Audioscrobbler 数据集中艺术家之间的余弦相似度	141
6.8	使用 Pig 以及 datafu 剔除 Audioscrobbler 数据集中的离群值	145

第 7 章　高级大数据分析 ··· 147

7.1	介绍	147
7.2	使用 Apache Giraph 计算 PageRank	147
7.3	使用 Apache Giraph 计算单源最短路径	150
7.4	使用 Apache Giraph 执行分布式宽度优先搜索	158
7.5	使用 Apache Mahout 计算协同过滤	165
7.6	使用 Apache Mahout 进行聚类	168
7.7	使用 Apache Mahout 进行情感分类	171

第 8 章　调试 ·· 174

8.1	介绍	174
8.2	在 MapReduce 中使用 Counters 监测异常记录	174
8.3	使用 MRUnit 开发和测试 MapReduce	177
8.4	本地模式下开发和测试 MapReduce	179
8.5	运行 MapReduce 作业跳过异常记录	182
8.6	在流计算作业中使用 Counters	184
8.7	更改任务状态显示调试信息	185
8.8	使用 illustrate 调试 Pig 作业	187

第 9 章　系统管理 ·· 189

9.1	介绍	189
9.2	在伪分布模式下启动 Hadoop	189

9.3 在分布式模式下启动 Hadoop ································· 192
9.4 添加一个新节点 ································· 195
9.5 节点安全退役 ································· 197
9.6 NameNode 故障恢复 ································· 198
9.7 使用 Ganglia 监控集群 ································· 199
9.8 MapReduce 作业参数调优 ································· 201

第 10 章 使用 Apache Accumulo 进行持久化 ································· 204

10.1 介绍 ································· 204
10.2 在 Accumulo 中设计行键存储地理事件 ································· 205
10.3 使用 MapReduce 批量导入地理事件数据到 Accumulo ································· 213
10.4 设置自定义字段约束 Accumulo 中的地理事件数据 ································· 220
10.5 使用正则过滤器限制查询结果 ································· 225
10.6 使用 SumCombiner 计算同一个键的不同版本的死亡数总和 ································· 228
10.7 使用 Accumulo 实行单元级安全的扫描 ································· 232
10.8 使用 MapReduce 聚集 Accumulo 中的消息源 ································· 237

第1章

Hadoop 分布式文件系统——导入和导出数据

本章我们将介绍：
- 使用 Hadoop shell 命令导入和导出数据到 HDFS
- 使用 `distcp` 实现集群间数据复制
- 使用 Sqoop 从 MySQL 数据库导入数据到 HDFS
- 使用 Sqoop 从 HDFS 导出数据到 MySQL
- 配置 Sqoop 以支持 SQL Server
- 从 HDFS 导出数据到 MongoDB
- 从 MongoDB 导入数据到 HDFS
- 使用 Pig 从 HDFS 导出数据到 MongoDB
- 在 Greenplum 外部表中使用 HDFS
- 利用 Flume 加载数据到 HDFS 中

1.1 介绍

在一个经典的数据架构中，Hadoop 是处理复杂数据流的核心。数据往往是从许多分散的系统中收集而来，并导入 **Hadoop 分布式文件系统（HDFS）** 中，然后通过 MapReduce 或者其他基于 MapReduce 封装的语言（如 Hive、Pig 和 Cascading 等）进行处理，最后将这些已经过滤、转换和聚合过的结果导出到一个或多个外部系统中。

举个比较具体的例子，一个大型网站可能会做一些关于网站点击率的基础数据分析。从多个服务器中采集页面访问日志，并将其推送到 HDFS 中。启动一个 MapReduce 作业，并将这些数据作为 MapReduce 的输入，接下来数据将被解析、汇总以及与 IP 地址进行关联计算，最终得出 URL、页面访问量和每个 cookie 的地理位置数据。生成的相关结果可以导入关系型数据库中。即席查询（Ad-hoc query）[①]此时就可以构建在这些数据上了。分析师可以快速地生成各种报表数据，例如，当前的独立用户数、用户访问最多的页面、按地区对用户进行拆分及其他的数据汇总。

本章的重点将关注 HDFS 数据的导入与导出，主要内容包含与本地文件系统、关系数据库、NoSQL 数据库、分布式数据库以及其他 Hadoop 集群之间数据的互相导入和导出。

1.2 使用 Hadoop shell 命令导入和导出数据到 HDFS

HDFS 提供了许多 shell 命令来实现访问文件系统的功能，这些命令都是构建在 HDFS FileSystem API 之上的。Hadoop 自带的 shell 脚本是通过命令行来执行所有操作的。这个脚本的名称叫做 hadoop，通常安装在$HADOOP_BIN 目录下，其中$HADOOP_BIN 是 Hadoopbin 文件完整的安装目录，同时有必要将$HADOOP_BIN 配置到$PATH 环境变量中，这样所有的命令都可以通过 `hadoop fs -command` 这样的形式来执行。

如果需要获取文件系统的所有命令，可以运行 hadoop 命令传递不带参数的选项 fs。

```
hadoop fs
```

```
[cloudera@localhost Desktop]$ hadoop fs
usage: hadoop fs [generic options]
    [-cat [-ignoreCrc] <src> ...]
    [-chgrp [-R] GROUP PATH...]
    [-chmod [-R] <MODE[,MODE]... | OCTALMODE> PATH...]
    [-chown [-R] [OWNER][:[GROUP]] PATH...]
    [-copyFromLocal <localsrc> ... <dst>]
    [-copyToLocal [-ignoreCrc] [-crc] <src> ... <localdst>]
    [-count [-q] <path> ...]
    [-cp <src> ... <dst>]
    [-df [-h] [<path> ...]]
    [-du [-s] [-h] <path> ...]
    [-expunge]
    [-get [-ignoreCrc] [-crc] <src> ... <localdst>]
    [-getmerge [-nl] <src> <localdst>]
    [-help [cmd ...]]
    [-ls [-d] [-h] [-R] [<path> ...]]
    [-mkdir [-p] <path> ...]
    [-moveFromLocal <localsrc> ... <dst>]
    [-moveToLocal <src> <localdst>]
    [-mv <src> ... <dst>]
    [-put <localsrc> ... <dst>]
    [-rm [-f] [-r|-R] [-skipTrash] <src> ...]
```

[①] 即席查询是用户根据自己的需求，灵活地选择查询条件，系统能够根据用户的选择生成相应的统计报表。

```
[-rmdir [--ignore-fail-on-non-empty] <dir> ...]
[-setrep [-R] [-w] <rep> <path/file> ...]
[-stat [format] <path> ...]
[-tail [-f] <file>]
[-test -[ezd] <path>]
[-text [-ignoreCrc] <src> ...]
[-touchz <path> ...]
[-usage [cmd ...]]
```

这些按照功能进行命名的命令的名称与 Unix shell 命令非常相似。使用 help 选项可以获得某个具体命令的详细说明。

```
hadoop fs -help ls
```

```
[cloudera@localhost Desktop]$ hadoop fs -help ls
-ls [-d] [-h] [-R] [<path> ...]:     List the contents that match the specified file pattern. If
        path is not specified, the contents of /user/currentUser
        will be listed. Directory entries are of the form
                dirName (full path) <dir>
        and file entries are of the form
                fileName(full path) <r n> size
        where n is the number of replicas specified for the file
        and size is the size of the file, in bytes.
        -d  Directories are listed as plain files.
        -h  Formats the sizes of files in a human-readable fashion
            rather than a number of bytes.
        -R  Recursively list the contents of directories.
[cloudera@localhost Desktop]$
```

这些 shell 命令和其简要的参数描述可在官方在线文档 http://hadoop.apache.org/docs/r1.0.4/file_system_shell.html 中进行查阅。

在这一节中，我们将使用 Hadoop shell 命令将数据导入 HDFS 中，以及将数据从 HDFS 中导出。这些命令更多地用于加载数据，下载处理过的数据，管理文件系统，以及预览相关数据。掌握这些命令是高效使用 HDFS 的前提。

准备工作

你需要在 http://www.packtpub.com/support 这个网站上下载数据集 weblog_entries.txt。

操作步骤

完成以下步骤，实现将 weblog_entries.txt 文件从本地文件系统复制到 HDFS 上的一个指定文件夹下。

1. 在 HDFS 中创建一个新文件夹，用于保存 weblog_entries.txt 文件：

```
hadoop fs -mkdir /data/weblogs
```

2. 将 weblog_entries.txt 文件从本地文件系统复制到 HDFS 刚创建的新文件夹下：

```
hadoop fs -copyFromLocal weblog_entries.txt /data/weblogs
```

3. 列出 HDFS 上 weblog_entries.txt 文件的信息：

```
hadoop fs -ls /data/weblogs/weblog_entries.txt
```

```
[cloudera@localhost Desktop]$ hadoop fs -ls /data/weblogs
Found 1 items
-rw-r--r--   1 cloudera supergroup     254129 2012-12-31 11:06 /data/weblogs/weblog_entries.txt
[cloudera@localhost Desktop]$
```

在Hadoop处理的一些结果数据可能会直接被外部系统使用,可能需要其他系统做更进一步的处理,或者MapReduce处理框架根本就不符合该场景,任何类似的情况下都需要从HDFS上导出数据。下载数据最简单的办法就是使用Hadoop shell。

4. 将HDFS上的 `weblog_entries.txt` 文件复制到本地系统的当前文件夹下:

`hadoop fs -copyToLocal /data/weblogs/weblog_entries.txt ./weblog_entries.txt`

```
[cloudera@localhost data]$ hadoop fs -copyToLocal /data/weblogs/weblog_entries.txt ./weblog_entries.txt
[cloudera@localhost data]$ ls -ltr
total 252
-rwxr-xr-x 1 cloudera cloudera 254129 Dec 31 11:15 weblog_entries.txt
[cloudera@localhost data]$
```

复制HDFS的文件到本地文件系统时,需要保证本地文件系统的空间可用以及网络连接的速度。HDFS中的文件大小在几个TB到几十个TB是很常见的。在1Gbit网络环境下,从HDFS中导出10 TB数据到本地文件系统,最好的情况下也要消耗23个小时,当然这还要保证本地文件系统的空间是可用的。

下载本书的示例代码

你可以从 http://www.packtpub.com 下载你买过的任何书的示例代码,如果你买过本书还可以访问 http://www.packtpub.com/support,并进行注册来让文件直接发送到你的邮箱

工作原理

Hadoop shell非常轻量地封装在HDFS FileSystem API之上。在执行 `hadoop` 命令时,如果传进去的参数是 `fs`,实际上执行的是 `org.apache.hadoop.fs.FsShell` 这个类。在0.20.2版本中 `FsShell` 实例化了一个 `org.apache.hadoop.fs.FileSystem` 对象,并且将命令行参数与类方法映射起来。比如,执行 `hadoop fs -mkdir /data/weblogs` 相当于调用 `FileSystem.mkdirs(new Path("/data/weblogs"))`。同样,运行 `hadoop fs -copyFromLocal weblog_entries.txt /data/weblogs` 相当于在调用 `FileSystem.copyFromLocal(newPath("weblog_entries.txt"), new Path("/data/weblogs"))`。HDFS数据复制到本地系统的实现方式也是一样,等同于调用 `FileSystem.copyToLocal(newPath("/data/weblogs/weblog_entries.txt"), new Path("./weblog_entries.txt"))`。更多关于文件系统

的接口信息描述可以见官方文档 http://hadoop.apache.org/docs/r1.0.4/api/org/apache/hadoop/fs/FileSystem.html。

  mkdir 可以通过 hadoop fs -mkdir PATH1 PATH2 的形式来执行。例如，hadoop fs -mkdir /data/weblogs/12012012 /data/weblogs/12022012 将会在 HDFS 系统上分别创建两个文件夹/data/weblogs/12012012 和/data/weblogs/12022012。如果文件夹创建成功则返回0，否则返回-1。

```
hadoop fs -mkdir /data/weblogs/12012012 /data/weblogs/12022012
hadoop fs -ls /data/weblogs
```

```
[cloudera@localhost data]$ hadoop fs -mkdir /data/weblogs/12012012 /data/weblogs/12022
012
[cloudera@localhost data]$ hadoop fs -ls /data/weblogs
Found 3 items
drwxr-xr-x   - cloudera supergroup          0 2012-12-31 11:18 /data/weblogs/12012012
drwxr-xr-x   - cloudera supergroup          0 2012-12-31 11:18 /data/weblogs/12022012
-rw-r--r--   1 cloudera supergroup     254129 2012-12-31 11:06 /data/weblogs/weblog_en
tries.txt
[cloudera@localhost data]$
```

  copyFromLocal 可以通过 hadoop fs -copyFromLocal LOCAL_FILE_PATH URI 的形式来执行，如果 URI 的地址（指的是 HDFS://filesystemName:9000 这个串）没有明确给出，则默认会读取 core-site.xml 中的 fs.default.name 这个属性。上传成功返回0，否则返回-1。

  copyToLocal 命令可以通过 hadoop fs -copyToLocal [-ignorecrc] [-crc] URI LOCAL_FILE_PATH 的形式来执行。如果 URI 的地址没有明确的给出，则默认会读取 core-site.xml 中的 fs.default.name 这个属性。copyToLocal 会执行 CRC（Cyclic Redundancy Check）校验保证已复制的数据的正确性，一个失败的副本复制可以通过参数-ignorecrc 来强制执行，还可以通过-crc 参数在复制文件的同时也复制 crc 校验文件。

更多参考

  put 命令与 copyFromLocal 类似，put 更通用一些，可以复制多个文件到 HDFS 中，也能接受标准输入流。

  任何使用 copyToLocal 的地方都可以用 get 替换，两者的实现一模一样。

  在使用 MapReduce 处理大数据时，其输出结果可能是一个或者多个文件。最终输出结果的文件个数是由 mapred.reduce.tasks 的值决定的。我们可以在 Jobconf 类中通过 setNumReduceTasks() 方法来设置这个属性，改变提交作业的 reduce 个数，每个 reduce 将对应输出一个文件。该参数是客户端参数，非集群参数，针对不同的作业应该设置不同的值。其默认值为1，意味着所有 Map（映射函数，以下都用 Map 表示）的输出结果都将复制到 1 个 reducer 上进行处理。除非 Map 输出的结果数据小于 1 GB，否则默认的配置将不合适。reduce 个数的设置更像是一门艺术而不是科学。在官方的文档中对其设置推荐的两个公式如下：

  0.95×NUMBER_OF_NODES×mapred.tasktracker.reduce.tasks.maximum

或者

  1.75×NUMBER_OF_NODES×mapred.tasktracker.reduce.tasks.maximum

  假设你的集群有 10 个节点来运行 Task Tracker，每个节点最多可以启动 5 个 reduce 槽位（通过设置 `tasktracker.reduce.tasks.maximum` 这个值来决定每个节点所能启动的最大 reduce 槽位数）对应的这个公式应该是 0.95×10×4=47.5。因为 reduce 个数的设置必须是整数，所以需要进行四舍五入。

  JobConf 文档中给出了使用这两个公式的理由，具体见 `http://hadoop.apache.org/docs/current/api/org/apache/hadoop/mapred/JobConf.html#setNumReduceTasks(int)`。

  0.95 可以保证在 map 结束后可以立即启动所有的 reduce 进行 map 结果的复制，只需要一波就可以完成作业。1.75 使得运行比较快的 reducer 能够再执行第二波 reduce，保证两波 reduce 就能完成作业，使作业整体的负载均衡保持得比较好。

  reduce 输出的数据存储在 HDFS 中，可以通过文件夹的名称来引用。若文件夹作为一个作业的输入，那么该文件夹下的所有文件都会被处理。上文介绍的 `get` 和 `copyToLocal` 只能对文件进行复制，无法对整个文件夹进行复制[①]。当然 Hadoop 提供了 `getmerge` 命令，可以将文件系统中的多个文件合并成一个单独的文件下载到本地文件系统。

  通过以下 Pig 脚本来演示下 getmerge 命令的功能：

```
weblogs = load '/data/weblogs/weblog_entries.txt' as
              (md5:chararray,
               url:chararray,
               date:chararray,
               time:chararray,
               ip:chararray);

md5_grp = group weblogs by md5 parallel 4;

store md5_grp into '/data/weblogs/weblogs_md5_groups.bcp';
```

  Pig 脚本可以通过下面的命令行来执行：

```
pig -f weblogs_md5_group.pig
```

---

① 原文是针对 Hadoop 0.20.0 的版本，对目前来说该版本已经很老，Hadoop 1.0 以上的版本 put 已经可以对文件夹进行复制。

```
[cloudera@localhost data]$ hadoop fs -ls /data/weblogs
Found 4 items
drwxr-xr-x   - cloudera supergroup          0 2012-12-31 11:18 /data/weblogs/12012012
drwxr-xr-x   - cloudera supergroup          0 2012-12-31 11:18 /data/weblogs/12022012
-rw-r--r--   1 cloudera supergroup     254129 2012-12-31 11:06 /data/weblogs/weblog_entries.txt
drwxr-xr-x   - cloudera supergroup          0 2012-12-31 11:27 /data/weblogs/weblogs_md5_groups.bcp
[cloudera@localhost data]$
```

该脚本逐行读取 HDFS 上的 weblog_entries.txt 文件，并且按照 md5 的值进行分组。parallel 是 Pig 脚本用来设置 reduce 个数的方法。由于启动了 4 个 reduce 任务，所以会在输出的目录 /data/weblogs/weblogs_md5_groups.bcp 中生成 4 个文件。

注意，weblogs_md5_groups.bcp 实际上是一个文件夹，显示该文件夹的列表信息可以看到：

```
[cloudera@localhost data]$ hadoop fs -ls /data/weblogs/weblogs_md5_groups.bcp
Found 5 items
-rw-r--r--   1 cloudera supergroup          0 2012-12-31 11:27 /data/weblogs/weblogs_md5_groups.bcp/_SUCCESS
-rw-r--r--   1 cloudera supergroup      85435 2012-12-31 11:27 /data/weblogs/weblogs_md5_groups.bcp/part-r-00000
-rw-r--r--   1 cloudera supergroup      91250 2012-12-31 11:27 /data/weblogs/weblogs_md5_groups.bcp/part-r-00001
-rw-r--r--   1 cloudera supergroup      87885 2012-12-31 11:27 /data/weblogs/weblogs_md5_groups.bcp/part-r-00002
-rw-r--r--   1 cloudera supergroup      90017 2012-12-31 11:27 /data/weblogs/weblogs_md5_groups.bcp/part-r-00003
[cloudera@localhost data]$
```

在 /data/weblogs/weblogs_md5_groups.bcp 中包含 4 个文件，即 part-r-00000、part-r-00001、part-r-00002 和 part-r-00003。

getmerge 命令可以用来将 4 个文件合并成一个文件，并且复制到本地的文件系统中，具体命令如下：

hadoop fs -getmerge /data/weblogs/weblogs_md5_groups.bcp weblogs_md5_groups.bcp

操作完我们可以看到本地文件列表如下：

```
[cloudera@localhost data]$ hadoop fs -getmerge /data/weblogs/weblogs_md5_groups.bcp weblogs_md5_groups.bcp
[cloudera@localhost data]$ ls -ltr
total 600
-rwxr-xr-x 1 cloudera cloudera 254129 Dec 31 11:15 weblog_entries.txt
-rwxr-xr-x 1 cloudera cloudera 359587 Dec 31 15:25 weblogs_md5_groups.bcp
[cloudera@localhost data]$
```

延伸阅读

- 关于 HDFS 数据的读写，我们将在第 2 章中重点介绍如何直接利用文件系统的 API 进行读写。

- 通过以下的两个链接可以对比出文件系统 shell 命令与 Java API 的不同：
http://hadoop.apache.org/docs/r1.0.4/file_system_shell.html
http://hadoop.apache.org/docs/r1.0.4/api/org/apache/hadoop/fs/FileSystem.html

1.3 使用 distcp 实现集群间数据复制

Hadoop 分布式复制（distcp）是 Hadoop 集群间复制大量数据的高效工具。distcp

是通过启动 MapReduce 实现数据复制的。使用 MapReduce 的好处包含可并行性、高容错性、作业恢复、日志记录、进度汇报等。Hadoop 分布式复制（`distcp`）对在开发集群环境、研究集群环境和生产集群环境之间进行数据复制十分有用。

准备工作

首先必须保证复制源和复制目的地能够互相访问。

最好关闭复制源集群 map 任务的推测机制，可以在配置文件 `mapred-site.xml` 中将 `mapred.map.tasks.speculative.execution` 的值设置为 `false` 来实现，这样就可以避免在 map 任务失败的时候产生任何不可知的行为。

源集群和目的集群的 RPC 协议必须是一致。这意味着两个集群之间安装的 Hadoop 版本必须一致[①]。

操作步骤

完成以下几个步骤实现集群间的文件夹复制。

1. 将集群 A 的 weblogs 文件夹复制到集群 B 上：

```
hadoop distcp hdfs://namenodeA/data/weblogs hdfs://namenodeB/data/weblogs
```

2. 将集群 A 的 weblogs 文件夹复制到集群 B 并覆盖已存在文件：

```
hadoop distcp -overwrite hdfs://namenodeA/data/weblogs hdfs://namenodeB/data/weblogs
```

3. 同步集群 A 和集群 B 之间的 weblogs 文件夹：

```
hadoop distcp -update hdfs://namenodeA/data/weblogs hdfs://namenodeB/data/weblogs
```

工作原理

在源集群，文件夹中的内容将被复制为一个临时的大文件。将会启动一个只有 map（map-only[②]）的 MapReduce 作业来实现两个集群间的数据复制。默认情况下，每个 map 就将会分配到一个 256 MB 的数据块文件。举个例子，如果 weblogs 文件夹总大小为 10 GB，默认将会启动 40 个 map，每个 map 会复制大约 256 MB 的数据。`distcp` 复制也可以通过参数手动设置启动的 map 数量。

```
hadoop distcp -m 10 hdfs://namenodeA/data/weblogs hdfs://namenodeB/data/ weblogs
```

---

① Hadoop 0.20.2 以上已经支持不同版本间的 `distcp` 复制了。——译者注
② "只有 map" 表示一个作业只启动 map 阶段没有启动 reduce 阶段。

在上面这个例子中，将会启动 10 个 map 进行数据复制。如果 weblogs 文件夹的总大小是 10 GB，那么每个 map 会复制大约 1 GB 的数据。

更多参考

如果要在运行的 Hadoop 版本不一样的两个集群之间进行数据复制，一般建议在复制源集群使用 `HftpFileSystem`[①]。`HftpFileSystem` 是一个只读的文件系统。相应的 distcp 命令只能在目标服务器上运行：

```
hadoop distcp hftp://namenodeA:port/data/weblogs hdfs://namenodeB/data/weblogs
```

在上面这条命令中，`port` 的值要与配置文件 `hdfs-site.xml` 中 `dfs.http.address` 属性的端口值一致。

1.4 使用 Sqoop 从 MySQL 数据库导入数据到 HDFS

Sqoop 是 Apache 基金下的一个项目，是庞大 Hadoop 生态圈中的一部分。在很多方面 Sqoop 和 `distcp` 很相似（见 1.3 节）。这两个工具都是构建在 MapReduce 之上的，利用了 MapReduce 的并行性和容错性。与集群间的数据复制不同，Sqoop 设计通过 JDBC 驱动连接实现 Hadoop 集群与关系数据库之间的数据复制。

它的功能非常广泛，本节将以网络日志条目为例展示如何使用 Sqoop 从 MySQL 数据库导入数据到 HDFS。

准备工作

本例子使用 Sqoop V1.3.0 版本。

如果你使用的是 CDH3 版本，Sqoop 默认是已经安装了。如果不是 CDH3，你可以通过 `https://ccp.cloudera.com/display/CDHDOC/Sqoop+Installation` 找到发行版的说明。

在本节假设你已经启动了一个 MySQL 实例，并且能够访问 Hadoop 集群[②]。`mysql.user` 该表配置了你运行 Sqoop 的那台机器上被允许连接的用户。访问 `http://dev.mysql.com/doc/refman/5.5/en/installing.html` 获取更多关于 MySQL 安装与配置的相关信息。

将 MySQL JDBC 驱动包复制到 `$SQOOP_HOME/libs`[③] 目录下。该驱动包可以从

[①] Hadoop ftp 文件系统。
[②] 保证 MySQL 与 Hadoop 集群中的每个节点间网络是相通的。
[③] $SQOOP_HOME 为 SQOOP 的安装目录。

http://dev.mysql.com/downloads/connector/j/下载。

操作步骤

完成以下步骤实现将 MySQL 表数据导出到 HDFS 中。

1. 在 MySQL 实例中创建一个新数据库：

```
CREATE DATABASE logs;
```

2. 创建并载入表 weblogs：

```
USE logs;
CREATE TABLE weblogs (
    md5             VARCHAR(32),
    url             VARCHAR(64),
    request_date    DATE,
    request_time    TIME,
    ip              VARCHAR(15)
);
LOAD DATA INFILE '/path/weblog_entries.txt' INTO TABLE weblogs
FIELDS TERMINATED BY '\t' LINES TERMINATED BY '\r\n';
```

3. 查询 weblogs 表的行数：

```
mysql> select count(*) from weblogs;
```

输出结果将会是：

```
+----------+
| count(*) |
+----------+
|     3000 |
+----------+
1 row in set (0.01 sec)
```

4. 将 MySQL 数据导出到 HDFS：

```
sqoop import -m 1 --connect jdbc:mysql://<HOST>:<PORT>/logs--username hdp_usr
--password test1 --table weblogs --target-dir /data/weblogs/import
```

输出结果将会是：

```
INFO orm.CompilationManager: Writing jar file:
/tmp/sqoop-jon/compile/f57ad8b208643698f3d01954eedb2e4d/weblogs.jar
WARN manager.MySQLManager: It looks like you are importing from mysql.
WARN manager.MySQLManager: This transfer can be faster! Use the --direct
WARN manager.MySQLManager: option to exercise a MySQL-specific fast path.
...
INFO mapred.JobClient: Map input records=3000
INFO mapred.JobClient: Spilled Records=0
INFO mapred.JobClient: Total committed heap usage (bytes)=85000192
INFO mapred.JobClient: Map output records=3000
INFO mapred.JobClient: SPLIT_RAW_BYTES=87
INFO mapreduce.ImportJobBase: Transferred 245.2451 KB in 13.7619 seconds
```

```
(17.8206 KB/sec)
INFO mapreduce.ImportJobBase: Retrieved 3000 records.
```

工作原理

Sqoop 连接数据库的 JDBC 驱动在 `--connect` 语句中定义，并从 `$SQOOP_HOME/libs` 目录中加载相应的包，其中 `$SQOOP_HOME` 为 Sqoop 安装的绝对路径。`--username` 和 `--password` 选项用于验证用户访问 MySQL 实例的权限。mysql.user 表必须包含 Hadoop 集群每个节点的主机域名以及相应的用户名，否则 Sqoop 将会抛出异常，表明相应的主机不允许被连接到 MySQL 服务器。

```
mysql> USE mysql;
mysql> select host, user from user;
```

显示输出如下：

```
+------------+------------+
| user       | host       |
+------------+------------+
| hdp_usr    | hdp01      |
| hdp_usr    | hdp02      |
| hdp_usr    | hdp03      |
| hdp_usr    | hdp04      |
| root       | 127.0.0.1  |
| root       | ::1        |
| root       | localhost  |
+------------+------------+
7 rows in set (1.04 sec)
```

在这个例子中，我们使用 `hdp_usr` 用户连接到 MySQL 服务器。我们的集群拥有 4 台机器，即 `hdp01`、`hdp02`、`hdp03` 和 `hdp04`。

`--table` 变量告诉 Sqoop 哪个表需要被导入。在我们的例子中，是要导入 weblogs 这个表到 HDFS。`--target-dir` 变量决定了导出的表数据将被存储在 HDFS 的哪个目录下：

```
hadoop fs -ls /data/weblogs/import
```

输出结果为：

```
-rw-r--r--   1 hdp_usr hdp_grp        0  2012-06-08  23:47 /data/
weblogs/import/_SUCCESS

drwxr-xr-x   - hdp_usr hdp_grp        0  2012-06-08  23:47 /data/
weblogs/import/_logs

-rw-r--r--   1 hdp_usr hdp_grp   251131  2012-06-08  23:47 /data/
weblogs/import/part-m-00000
```

默认情况下，导入的数据将按主键进行分割。如果导入的表并不包含主键，必须指定 `-m` 或者 `--split-by` 参数决定导入的数据如何分割。在前面的例子中，使用了 `-m` 参数。`-m` 参数决定了将会启动多少个 mapper 来执行数据导入。因为将 `-m` 设置为 1，所以就启动了一个 mapper 用于导入数据。每个 mapper 将产生一个独立的文件。

这行命令背后隐藏了相当复杂的逻辑。Sqoop 利用数据库中存储的元数据生成每一列的 `DBWritable` 类，这些类使用了 `DBInputFormat`。`DBInputFormat` 是 Hadoop 用来格式化读取数据库任意查询的结果。在前面的例子中，启动了一个使用 `DBInputFormat` 索引 weblogs 表内容的 MapReduce 作业。整个 weblogs 表被扫描并存储在 HDFS 的路径 `/data/weblogs/import` 下。

更多参考

使用 Sqoop 导入数据还有很多有用的参数可以配置。Sqoop 可以分别通过参数 `--as-avrodatafile` 和 `--as-sequencefile` 将数据导入为 Avro 文件和序列化的文件。通过 `-z` 或者 `--compress` 参数可以在导入的过程中对数据进行压缩。默认的压缩方式为 GZIP 压缩，可以通过 `--compression-codec <CODEC>` 参数使用 Hadoop 支持的任何压缩编码。可以查看第 2 章的使用 LZO 压缩数据那一节的介绍。另一个有用的参数是 `--direct`，该参数指示 Sqoop 直接使用数据库支持的本地导入导出工具。在前面的例子中，如果 `--direct` 被添加为参数，Sqoop 将使用 `mysqldump` 工具更快地导出 weblogs 表的数据。`--direct` 参数非常重要以至于我们在运行前面的日志会打印出如下的日志信息：

```
WARN manager.MySQLManager: It looks like you are importing from mysql.
WARN manager.MySQLManager: This transfer can be faster! Use the --direct
WARN manager.MySQLManager: option to exercise a MySQL-specific fast path.
```

延伸阅读

❑ 使用 Sqoop 从 HDFS 导出数据到 MySQL（1.5 节）。

1.5 使用 Sqoop 从 HDFS 导出数据到 MySQL

Sqoop 是 Apache 基金会下的一个项目，是庞大 Hadoop 生态圈中的一部分。在很多方面 Sqoop 和 distcp 很相似（见 1.3 节）。这两个工具都是构建在 MapReduce 之上的，利用了 MapReduce 的并行性和容错性。与集群间的数据复制不同，Sqoop 设计通过 JDBC 驱动连接实现 Hadoop 集群与关系数据库之间的数据复制。

它的功能非常广泛，本节将以网络日志条目为例展示如何使用 Sqoop 从 HDFS 导入数据到 MySQL 数据库。

准备工作

本例使用 Sqoop V1.3.0 版本。

如果你使用的是 CDH3 版本，Sqoop 默认是已经安装了。如果不是 CDH3，你可以通过 `https://ccp.cloudera.com/display/CDHDOC/Sqoop+Installation` 找到发行版的说明。

在本节假设你已经启动了一个 MySQL 实例，并且能够访问 Hadoop 集群。mysql.user 表配置了你运行 Sqoop 的那台机器上被允许连接的用户。访问 `http://dev.mysql.com/doc/refman/5.5/en/installing.html` 获取更多关于 MySQL 安装与配置的相关信息。

将 MySQL JDBC 驱动包复制到 `$SQOOP_HOME/libs` 目录下。该驱动包可以从 `http://dev.mysql.com/downloads/connector/j/` 下载。

按照 1.1 节介绍的导入 `weblog_entires.txt` 文件到 HDFS 的方式操作。

操作步骤

完成以下步骤实现将 HDFS 数据导出到 MySQL 表中。

1. 在 MySQL 实例中创建一个新数据库：

```
CREATE DATABASE logs;
```

2. 创建表 `weblogs_from_hdfs`：

```
USE logs;
CREATE TABLE weblogs_from_hdfs (
    md5             VARCHAR(32),
    url             VARCHAR(64),
    request_date    DATE,
    request_time    TIME,
    ip              VARCHAR(15)
);
```

3. 从 HDFS 导出 `weblog_entries.txt` 文件到 MySQL：

```
sqoop export -m 1 --connect jdbc:mysql://<HOST>:<PORT>/logs --username hdp_usr
--password test1 --table weblogs_from_hdfs --export-dir /data/weblogs/05102012
--input-fields-terminated-by '\t' --mysql-delmiters
```

输出结果如下：

```
INFO mapreduce.ExportJobBase: Beginning export of weblogs_from_
hdfs
input.FileInputFormat: Total input paths to process : 1
input.FileInputFormat: Total input paths to process : 1
mapred.JobClient: Running job: job_201206222224_9010
INFO mapred.JobClient: Map-Reduce Framework
INFO mapred.JobClient: Map input records=3000
INFO mapred.JobClient: Spilled Records=0
INFO mapred.JobClient: Total committed heap usage
(bytes)=85000192
INFO mapred.JobClient: Map output records=3000
```

```
INFO mapred.JobClient: SPLIT_RAW_BYTES=133
INFO mapreduce.ExportJobBase: Transferred 248.3086 KB in 12.2398
seconds (20.287 KB/sec)
INFO mapreduce.ExportJobBase: Exported 3000 records.
```

工作原理

Sqoop 连接数据库的 JDBC 驱动可使用 -connect 参数声明,并从 `$SQOOP_HOME/libs` 目录中加载相应的包。其中 `$SQOOP_HOME` 为 Sqoop 安装的绝对路径。--username 和 --password 选项用于验证用户访问 MySQL 实例的权限。mysql.user 表必须包含 Hadoop 集群每个节点的主机域名以及相应的用户名,否则 Sqoop 将会抛出异常,表明相应的主机不允许被连接到 MySQL 服务器。

```
mysql> USE mysql;
mysql> select host, user from user;

+----------------+-----------+
| user           | host      |
+----------------+-----------+
| hdp_usr        | hdp01     |
| hdp_usr        | hdp02     |
| hdp_usr        | hdp03     |
| hdp_usr        | hdp04     |
| root           | 127.0.0.1 |
| root           | ::1       |
| root           | localhost |
+----------------+-----------+
7 rows in set (1.04 sec)
```

在这个例子中,我们使用 hdp_usr 用户连接到 MySQL 服务器。我们的集群拥有 4 台机器,即 hdp01、hdp02、hdp03 和 hdp04。

--table 参数决定了 HDFS 导出的数据将存储在哪个 MySQL 表中。这个表必须在执行 Sqoop export 语句之前创建好。Sqoop 通过表的元数据信息、列数量以及列类型来校验 HDFS 需要导出目录中的数据并生成相应的插入语句。举个例子,导出作业可以被想象为逐行读取 HDFS 的 weblogs_entries.txt 文件并产生以下输出:

```
INSERT INTO weblogs_from_hdfs
VALUES('aabba15edcd0c8042a14bf216c5', '/jcwbtvnkkujo.html', '2012-05- 10',
'21:25:44', '148.113.13.214');

INSERT INTO weblogs_from_hdfs
VALUES('e7d3f242f111c1b522137481d8508ab7', '/ckyhatbpxu.html', '2012- 05-10',
'21:11:20', '4.175.198.160');

INSERT INTO weblogs_from_hdfs
VALUES('b8bd62a5c4ede37b9e77893e043fc1', '/rr.html', '2012-05-10', '21:32:08',
'24.146.153.181');
...
```

Sqoop export 默认情况下是创建新增语句。如果--update-key 参数被设置了，则将是创建更新语句。如果前面的例子使用了参数--update-key md5 那么生成的 Sql 代码将运行如下：

```
UPDATE weblogs_from_hdfs SET url='/jcwbtvnkkujo.html', request_ date='2012-
05-10'request_time='21:25:44'
ip='148.113.13.214'WHERE md5='aabba15edcd0c8042a14bf216c5'

UPDATE weblogs_from_hdfs SET url='/jcwbtvnkkujo.html', request_ date='2012-05-
10' request_time='21:11:20' ip='4.175.198.160' WHERE md5='e7d3f242f111c1b
522137481d8508ab7'

UPDATE weblogs_from_hdfs SET url='/jcwbtvnkkujo.html', request_ date='2012-
05-10'request_time='21:32:08' ip='24.146.153.181' WHERE md5='b8bd62a5c4ede37b
9e77893e043fc1'
```

如果--update-key 设置的值并没找到，可以设置--update-mode 为 allowinsert 允许新增这行数据。

-m 参数决定将配置几个 mapper 来读取 HDFS 上文件块。每个 mapper 各自建立与 MySQL 服务器的连接。每个语句将会插入 100 条记录。当完成 100 条语句也就是插入 10000 条记录，将会提交当前事务。一个失败的 map 任务，很可能导致数据的不一致，从而出现插入冲突数据或者插入重复数据。这种情况可以通过使用参数--staging-table 来解决。这会促使任务将数据插入一个临时表，等待一个事务完成再将数据从临时表复制到 --table 参数配置的表中。临时表结构必须与最终表一致。临时表必须是一个空表否则需要配置参数--clear-staging-table。

延伸阅读

- 使用 Sqoop 从 MySQL 数据库导入 HDFS（1.4 节）。

1.6 配置 Sqoop 以支持 SQL Server

本节将展示如何配置 Sqoop 和 SQL Server 数据库进行连接。这样可以允许数据从 SQL Server 数据库有效地导入 HDFS 中。

准备工作

本例子使用了 Sqoop V1.3.0 版本。

如果你使用的是 CDH3 版本，Sqoop 默认是已经安装了。如果不是 CDH3，你可以通过 https://ccp.cloudera.com/display/CDHDOC/Sqoop+Installation 找到发行版的说明。

在本节假设你已经启动了一个 Microsoft SQL Server 实例，并且能够与 Hadoop 集群正常连接。

操作步骤

完成以下步骤配置 Sqoop 连接 Microsoft SQL Server。

1. 从 `http://download.microsoft.com/download/D/6/A/D6A241AC-433E-4CD2-A1CE- 50177E8428F0/1033/sqljdbc_3.0.1301.101_enu.tar.gz` 下载 Microsoft SQL Server JDBC 驱动 3.0。该下载包包含了 SQL Server JDBC 驱动（`sqljdbc4.jar`）。Sqoop 连接关系数据库是通过 JDBC 驱动的。

2. 解压缩 TAR 文件：
```
gzip -d sqljdbc_3.0.1301.101_enu.tar.gz
tar -xvf sqljdbc_3.0.1301.101_enu.tar
```
这将创建一个新的文件夹 `sqljdbc_3.0`。

3. 复制 `sqljdbc4.jar` 到`$SQOOP_HOME/lib` 目录下：
```
cp sqljdbc_3.0/enu/sqljdbc4.jar $SQOOP_HOME/lib
```
Sqoop 可以访问 `sqljdbc4.jar` 文件，并且可以通过该驱动包访问 SQL Server 实例。

4. 为 Hadoop 下载微软 SQL Server 连接器：
```
http://download.microsoft.com/download/B/E/5/BE5EC4FD-9EDA-4C3F-8B36-1C8AC4CE2CEF/sqoop-sqlserver-1.0.tar.gz
```

5. 解压缩 TAR 文件：
```
gzip -d sqoop-sqlserver-1.0.tar.gz
tar -xvf sqoop-sqlserver-1.0.tar
```
这将创建一个新的文件夹 `sqoop-sqlserver-1.0`。

6. 设置 `MSSQL_CONNECTOR_HOME` 环境变量：
```
export MSSQL_CONNECTOR_HOME=/path/to/sqoop-sqlserver-1.0
```

7. 运行安装脚本：
```
./install.sh
```

8. 导入导出数据可以查看 1.4 节和 1.5 节。这两节的内容对 SQL Server 同样适用，只是需要把--connect 参数修改为–connect `jdbc:sqlserver://<HOST>:<PORT>`。

工作原理

Sqoop 和数据库之间通过 JDBC 连接。只要将 `sqljdbc4.jar` 添加到`$SQOOP_HOME/lib`

目录下，Sqoop 就可以使用 `--connect jdbc:sqlserver://<HOST>:<PORT>` 连接 SQL Server 数据库实例。为了使 SQL Server 与 Sqoop 有充分的兼容性，修改了一些必要的配置，这些配置可以通过运行 `install.sh` 脚本来实现更新。

1.7 从 HDFS 导出数据到 MongoDB

本节将使用 `MongoOutputFormat` 类加载 HDFS 中的数据并收集到 MongoDB 中。

准备工作

使用 Mongo Hadoop 适配器最简单的方法是从 GitHub 上克隆 Mongo-Hadoop 工程，并且将该工程编译到一个特定的 Hadoop 版本。克隆该工程需要安装一个 Git 客户端。

本节假定你使用的 Hadoop 版本是 CDH3。

Git 客户端官方的下载地址是：`http://git-scm.com/downloads`。

在 Windows 操作系统上可以通过 `http://windows.github.com/` 访问 GitHub。

在 Mac 操作系统上可以通过 `http://mac.github.com/` 访问 GitHub。

可以通过 `https://github.com/mongodb/mongo-hadoop` 获取到 Mongo Hadoop 适配器。该工程需要编译在特定的 Hadoop 版本上。编译完的 JAR 文件需要复制到 Hadoop 集群每个节点的 `$HADOOP_HOME/lib` 目录下。

Mongo Java 的驱动包也需要安装到 Hadoop 集群每个节点的 `$HADOOP_HOME/lib` 目录下。该驱动包可从 `https://github.com/mongodb/mongo-java-driver/downloads` 下载。

操作步骤

完成下面步骤实现将 HDFS 数据复制到 MongoDB。

1. 通过下面的命令实现克隆 mongo-hadoop 工程：
```
git clone https://github.com/mongodb/mongo-hadoop.git
```

2. 切换到稳定发布的 1.0 分支版本：
```
git checkout release-1.0
```

3. 必须保持 mongo-hadoop 与 Hadoop 的版本一致。使用文本编辑器打开 mongo-hadoop 克隆目录下的 `build.sbt` 文件，将下面这行：
```
hadoopRelease in ThisBuild := "default"
```
修改为：

```
hadoopRelease in ThisBuild := "cdh3"
```

4. 编译 `mongo-hadoop`：

```
./sbt package.
```

这将会在 `core/target` 文件夹下生成一个名为 `mongo-hadoop-core_cdh3u3-1.0.0.jar` 的 JAR 文件。

5. 从 https://github.com/mongodb/mongo-java-driver/downloads 下载 MongoDB 2.8.0 版本的 Java 驱动包。

6. 复制 `mongo-hadoop` 和 MongoDB Java 驱动包到 Hadoop 集群每个节点的 `$HADOOP_HOME/lib`：

```
cp mongo-hadoop-core_cdh3u3-1.0.0.jar mongo-2.8.0.jar $HADOOP_HOME/lib
```

7. 编写 MapReduce 读取 HDFS 上 `weblog_entries.txt` 文件并通过 `MongoOutputFormat` 类将数据写入 MongoDB 中：

```java
import java.io.*;

import org.apache.commons.logging.*;
import org.apache.hadoop.conf.*;
import org.apache.hadoop.fs.Path;
import org.apache.hadoop.io.*;
import org.apache.hadoop.mapreduce.lib.input.FileInputFormat;
import org.apache.hadoop.mapreduce.lib.input.TextInputFormat;
import org.apache.hadoop.mapreduce.*;
import org.bson.*;
import org.bson.types.ObjectId;

import com.mongodb.hadoop.*;
import com.mongodb.hadoop.util.*;

public class ExportToMongoDBFromHDFS {

    private static final Log log =
LogFactory.getLog(ExportToMongoDBFromHDFS.class);
    public static class ReadWeblogs extends Mapper<LongWritable, Text, ObjectId, BSONObject>{

        public void map(Text key, Text value, Context context) throws IOException, InterruptedException{

            System.out.println("Key: " + key);
            System.out.println("Value: " + value);

            String[] fields = value.toString().split("\t");

            String md5 = fields[0];
            String url = fields[1];
            String date = fields[2];
```

```java
        String time = fields[3];
        String ip = fields[4];

        BSONObject b = new BasicBSONObject();
        b.put("md5", md5);
        b.put("url", url);
        b.put("date", date);
        b.put("time", time);
        b.put("ip", ip);

        context.write( new ObjectId(), b);
    }
}

    public static void main(String[] args) throws Exception{

        final Configuration conf = new Configuration();
        MongoConfigUtil.setOutputURI(conf,
        "mongodb://<HOST>:<PORT>/test. weblogs");

        System.out.println("Configuration: " + conf);

        final Job job = new Job(conf, "Export to Mongo");

        Path in = new Path("/data/weblogs/weblog_entries.txt");
        FileInputFormat.setInputPaths(job, in);

        job.setJarByClass(ExportToMongoDBFromHDFS.class);
        job.setMapperClass(ReadWeblogs.class);

        job.setOutputKeyClass(ObjectId.class);
        job.setOutputValueClass(BSONObject.class);

        job.setInputFormatClass(TextInputFormat.class);
        job.setOutputFormatClass(MongoOutputFormat.class);

        job.setNumReduceTasks(0);

        System.exit(job.waitForCompletion(true) ? 0 : 1 );
    }
}
```

8. 导出为一个可运行的 JAR 文件，并运行该作业：

```
hadoop jar ExportToMongoDBFromHDFS.jar
```

9. 在 Mongo shell 上验证 weblogs 已经导入 MongoDB：

```
db.weblogs.find();
```

工作原理

Mongo Hadoop 适配器提供了一种新的兼容 Hadoop 的文件系统实现包括 `MongoInputFormat`

和 `MongoOutputFormat`。这些抽象实现使得访问 MongoDB 和访问任何兼容 Hadoop 的文件系统一样。

1.8 从 MongoDB 导入数据到 HDFS

本节将使用 `MongoInputFormat` 类加载 MongoDB 中的数据导入 HDFS 中。

准备工作

使用 Mongo Hadoop 适配器最简单的方法是从 GitHub 上克隆 Mongo-Hadoop 工程，并且将该工程编译到一个特定的 Hadoop 版本。克隆该工程需要安装一个 Git 客户端。

本节假定你使用的 Hadoop 版本是 CDH3。

Git 客户端官方的下载地址是：http://git-scm.com/downloads。

在 Windows 操作系统上可以通过 http://windows.github.com/ 访问 GitHub。

在 Mac 操作系统上可以通过 http://mac.github.com/ 访问 GitHub。

可以通过 https://github.com/mongodb/mongo-hadoop 获取到 Mongo Hadoop 适配器。该工程需要编译在特定的 Hadoop 版本上。编译完的 JAR 文件需要复制到 Hadoop 集群每个节点的 `$HADOOP_HOME/lib` 目录下。

Mongo Java 驱动包也需要安装到 Hadoop 集群每个节点的 `$HADOOP_HOME/lib` 目录下。该驱动包可从 https://github.com/mongodb/mongo-java-driver/downloads 下载。

操作步骤

完成下面步骤实现将 MongoDB 中的数据复制到 HDFS 中。

1. 通过下面的命令实现克隆 mongo-hadoop 工程：

`git clone https://github.com/mongodb/mongo-hadoop.git`

2. 切换到稳定发布的 1.0 分支版本：

`git checkout release-1.0`

3. 必须保持 `mongo-hadoop` 与 Hadoop 的版本一致。使用文本编辑器打开 `mongo-hadoop` 克隆目录下的 `build.sbt` 文件，修改下面这行：

`hadoopRelease in ThisBuild := "default"`

修改为：

```
hadoopRelease in ThisBuild := "cdh3"
```

4. 编译 `mongo-hadoop`：

```
./sbt package.
```

这将会在 `core/target` 文件夹下生成一个名为 `mongo-hadoop-core_cdh3u3-1.0.0.jar` 的 JAR 文件。

5. 从 `https://github.com/mongodb/mongo-java-driver/downloads` 下载 MongoDB 2.8.0 版本的 Java 驱动包。

6. 复制 `mongo-hadoop` 和 MongoDB Java 驱动包到 Hadoop 集群每个节点的 `$HADOOP_HOME/lib`：

```
cp mongo-hadoop-core_cdh3u3-1.0.0.jar mongo-2.8.0.jar $HADOOP_HOME/lib
```

7. 编写 MapReduce 读取 MongoDB 数据库中的数据并写入 HDFS 中：

```java
import java.io.*;

import org.apache.commons.logging.*;
import org.apache.hadoop.conf.*;
import org.apache.hadoop.fs.Path;
import org.apache.hadoop.io.*;
import org.apache.hadoop.mapreduce.lib.output.*;
import org.apache.hadoop.mapreduce.*;
import org.bson.*;

import com.mongodb.hadoop.*;
import com.mongodb.hadoop.util.*;

public class ImportWeblogsFromMongo {

    private static final Log log = LogFactory.getLog(ImportWeblogsFrom Mongo.class);

    public static class ReadWeblogsFromMongo extends Mapper<Object, BSONObject, Text, Text>{

        public void map(Object key, BSONObject value, Context context) throws IOException, InterruptedException{
            System.out.println("Key: " + key);
            System.out.println("Value: " + value);

            String md5 = value.get("md5").toString();
            String url = value.get("url").toString();
            String date = value.get("date").toString();
            String time = value.get("time").toString();
            String ip = value.get("ip").toString();
            String output = "\t" + url + "\t" + date + "\t" +
                            time + "\t" + ip;
```

```
            context.write( new Text(md5), new Text(output));
        }
    }

    public static void main(String[] args) throws Exception{

        final Configuration conf = new Configuration();
        MongoConfigUtil.setInputURI(conf, "mongodb://<HOST>:<PORT>/test.weblogs");
        MongoConfigUtil.setCreateInputSplits(conf, false);
        System.out.println("Configuration: " + conf);

        final Job job = new Job(conf, "Mongo Import");

        Path out = new Path("/data/weblogs/mongo_import");
        FileOutputFormat.setOutputPath(job, out);
        job.setJarByClass(ImportWeblogsFromMongo.class);
        job.setMapperClass(ReadWeblogsFromMongo.class);
        job.setOutputKeyClass(Text.class);
        job.setOutputValueClass(Text.class);

        job.setInputFormatClass(MongoInputFormat.class);
        job.setOutputFormatClass(TextOutputFormat.class);

        job.setNumReduceTasks(0);

        System.exit(job.waitForCompletion(true) ? 0 : 1 );

    }
}
```

这个只有 map 的作业用到了 Mongo Hadoop 适配器提供的几个类。从 HDFS 读入的数据会被转换成一个 `BSONObject` 对象。该类描述的是一个二进制的 JSON 值。MongoDB 使用这些 `BSONObject` 对象来有效地序列化、传输和存储数据。

Mongo Hadoop 适配器还提供了一个方便的工具类 `MongoConfigUtil`，使得可以把 MongoDB 当成是一个文件系统来访问。

8. 导出为一个可运行的 JAR 文件，并运行该作业：

```
hadoop jar ImportWeblogsFromMongo.jar
```

9. 验证 weblogs 数据是否已经导入 HDFS 中：

```
hadoop fs -ls /data/weblogs/mongo_import
```

工作原理

Mongo Hadoop 适配器提供了一种新的兼容 Hadoop 的文件系统实现，包括 `MongoInputFormat` 和 `MongoOutputFormat`。这些抽象实现使得访问 MongoDB 和访问任何兼容 Hadoop 的文件系统一样。

1.9 使用 Pig 从 HDFS 导出数据到 MongoDB

MongoDB 是一种 NoSQL 数据库，用于存储和检索海量数据。MongoDB 通常用于存储面向用户的数据，这些数据必须经过清洗、格式化之后才可以被使用。Apache Pig 从某种程度上讲就是用来处理这种任务的。`Mongostorage` 类使得使用 Pig 可以非常方便地批量处理 HDFS 上的数据，再直接将这些数据导入 MongoDB 中。本节将使用 `Mongostorage` 类将 HDFS 上的数据导出到 MongoDB 数据库中。

准备工作

使用 Mongo Hadoop 适配器最简单的方法是从 GitHub 上克隆 `Mongo-Hadoop` 工程，并且将该工程编译到一个特定的 Hadoop 版本。克隆该工程需要安装一个 Git 客户端。

本节假定你使用的 Hadoop 版本是 CDH3。

Git 客户端官方的下载地址是：`http://git-scm.com/downloads`。

在 Windows 操作系统上可以通过 `http://windows.github.com/` 访问 GitHub。

在 Mac 操作系统上可以通过 `http://mac.github.com/` 访问 GitHub。

可以通过 `https://github.com/mongodb/mongo-hadoop` 获取到 Mongo Hadoop 适配器。该工程需要编译在特定的 Hadoop 版本上。编译完的 JAR 文件需要复制到 Hadoop 集群每个节点的 `$HADOOP_HOME/lib` 目录下。

Mongo Java 的驱动包也需要安装到 Hadoop 集群每个节点的 `$HADOOP_HOME/lib` 目录下。该驱动包可从 `https://github.com/mongodb/mongo-java-driver/downloads` 下载。

操作步骤

完成下面的步骤，将数据从 HDFS 复制到 MongoDB。

1. 通过下面的命令实现克隆 `mongo-hadoop` 工程：

```
git clone https://github.com/mongodb/mongo-hadoop.git
```

2. 切换到稳定发布的 1.0 分支版本：

```
git checkout release-1.0
```

3. 必须保持 `mongo-hadoop` 与 Hadoop 的版本一致。使用文本编辑器打开 `mongo-hadoop` 克隆目录下的 `build.sbt` 文件，修改下面这行：

```
hadoopRelease in ThisBuild := "default"
```

修改为：

```
hadoopRelease in ThisBuild := "cdh3"
```

4. 编译 `mongo-hadoop`：

`./sbt package`.

这将会在 `core/target` 文件夹下生成一个名为 `mongo-hadoop-core_cdh3u3-1.0.0.jar` 的 JAR 文件。

5. 从 `https://github.com/mongodb/mongo-java-driver/downloads` 下载 MongoDB 2.8.0 版本的 Java 驱动包。

6. 复制 `mongo-hadoop-core`、`mongo-hadoop-pig` 和 MongoDB Java 驱动包到 Hadoop 集群每个节点的 `$HADOOP_HOME/lib`：

```
cp mongo-hadoop-core_cdh3u3-1.0.0.jar mongo-2.8.0.jar $HADOOP_HOME/lib
```

7. 创建一个 Pig 脚本读取 HDFS 上的 weblogs 数据并将其存储到 MongoDB 数据库：

```
register /path/to/mongo-hadoop/mongo-2.8.0.jar
register /path/to/mongo-hadoop/core/target/mongo-hadoop-core- 1.0.0.jar
register /path/to/mongo-hadoop/pig/target/mongo-hadoop-pig- 1.0.0.jar

define MongoStorage com.mongodb.hadoop.pig.MongoStorage();

weblogs = load '/data/weblogs/weblog_entries.txt' as
              (md5:chararray, url:chararry, date:chararray, time:chararray,
               ip:chararray);

store weblogs into 'mongodb://<HOST>:<PORT>/test.weblogs_from_pig' using
MongoStorage();
```

工作原理

Mongo Hadoop 适配器提供了一种新的兼容 Hadoop 的文件系统实现，包括 `MongoInputFormat` 和 `MongoOutputFormat`。这些抽象实现使得访问 MongoDB 和访问任何兼容 Hadoop 的文件系统一样。`MongoStorage` 将 Pig 类型转化为 MongoDB 可以直接访问的 `BasicDBObjectBuilder` 类型。

1.10 在 Greenplum 外部表中使用 HDFS

Greenplum 是一个并行数据库，数据的存储与查询基于一个或多个 PostgreSQL 实例。它补充了 Hadoop，提供对大数据的实时或准实时访问，它还支持使用 HDFS 文件作为外部表。外部表是一个处理 Greenplum 集群之外数据很好的解决方案。由于外部表访问首先要

Hadoop 分布式文件系统——导入和导出数据

消耗网络带宽,所以与 Greenplum 集群内的数据相比,它应该存储那些访问相对不频繁的数据。本节将介绍如何创建只读的外部表和可读写的外部表。

准备工作

在本节假定你使用的 Hadoop 版本为 CDH3。

运行一个 Greenplum 实例,确认它能够访问 Hadoop 集群。相关信息可以查看 http://www.greenplum.com/products/greenplum-database。

按照下面的方式配置 Greenplum。

- gp_hadoop_target_version 设置为 cdh3u2。
- gp_hadoop_home 设置为$HADOOP_HOME 的完整路径。

在 Greenplum 集群的每个节点上都需要安装 1.6 以上(包含 1.6)版本的 Java。

操作步骤

创建 HDFS 上 weblogs 文件的一个外部表:

```
CREATE EXTERNAL TABLE weblogs(
    md5             text,
    url             text,
    request_date    date,
    request_time    time,
    ip              inet
)
LOCATION ('gphdfs://<NAMENODE_HOST>:<NAMENODE_PORT>/data/weblogs/weblog_ entries.txt')
FORMAT 'TEXT' (DELIMITER '\t');
```

工作原理

Greenplum 本地支持并行地将 HDFS 上的数据加载到数据库中。当一个查询需要访问表 weblog_entries.txt 时,该文件会被加载为 Greenplum 的一个临时表,然后执行对该临时表的查询。等到查询结束以后再将该表丢弃。

更多参考

Greenplum 也支持对外部表的写操作。这需要在创建表的时候指定写关键字:

```
CREATE WRITABLE EXTERNAL TABLE weblogs(
    md5             text,
    url             text,
    request_date    date,
    request_time    time,
    ip              inet
)
```

```
LOCATION ('gphdfs://<NAMENODE_HOST>:<NAMENODE_PORT>/data/weblogs/
weblog_ entries.txt')
FORMAT 'TEXT' (DELIMITER '\t');
```

更多的信息可以查看 Greenplum 管理员手册，见 http://media.gpadmin.me/wp-content/uploads/2011/05/GP-4100-AdminGuide.pdf。

1.11 利用 Flume 加载数据到 HDFS 中

Apache Flume 是 Hadoop 社区的一个项目，由多个相关项目组成，用于从不同的数据源可靠有效地加载数据流到 HDFS 中。Flume 最常见的一个场景是加载多个数据源的网站日志数据。本节将介绍如何使用 Flume 加载数据到 HDFS 中。

准备工作

在本节中假定你已经安装和配置好 Flume。

Flume 可以从 Apache 网页（http://incubator.apache.org/flume/）下载。

如果你使用的是 CDH3，那么默认已经安装了 Flume 0.9.4+25.43 的版本。

操作步骤

完成下面的步骤，实现将 weblogs 数据导入 HDFS。

1. 使用 dump 命令测试 Flume 是否配置正确：

```
flume dump 'text("/path/to/weblog_entries.txt")'
```

2. 通过 Flume shell 执行一个配置：

```
flume shell -c<MASTER_HOST>:<MASTER_PORT> -e 'exec config text("/path/to/
weblog_entries.txt") | collectorSink("hdfs://<NAMENODE_HOST>:<NAMENODE_PORT>/
data/weblogs/flume")'
```

工作原理

Flume 包含 Sources[①] 和 Sinks[②] 两个抽象概念，并通过管状的数据流将它们合并在一起。在这个例子中，数据来源是 text 方式，将文件路径作为参数，并将该文件中的内容发送到配置的数据输出端。dump 命令使用控制台为数据输出端。按照这样的配置，weblog_entries.txt 文件的内容以 text 的方式被读取，同时被写到控制台。

---

① 数据来源。
② 数据去向。

在步骤 2 中，使用了 Flume shell 配置并执行一个作业。-c 参数告诉了 Flume Master 节点的连接地址。Flume 将会执行-e 参数后面的命令。如前所述，text 是一种读取所传文件中所有内容的数据源。collectorSink 是一个数据去向，可以传给本地文件系统路径或者 HDFS 文件系统路径。在前面这个例子中，我们传递的是一个 HDFS 的路径。这个命令执行的效果会将 weblog_entries.txt 导入 HDFS 中。

更多参考

Flume 提供了几个预定义的 Source 和 Sink。下面是一些基本的数据源。

- null：不读取任何数据。
- stdin：读入一个标准的输入流。
- rpcSource：读取 Thrift 或者 Avro RPC。
- text：读取一个文件中的内容。
- tail：读取一个文件，并保持文件打开状态用于持续读取追加到文件中的数据

下面是一些基本的 Sink。

- null：将数据丢弃，不进行写操作。
- collectorSink：写到一个本地文件系统或者 HDFS 上。
- console：写到控制台。
- formatDfs：写到 HDFS 并带一定的格式，如序列化文件、Avro、Thrift 等。
- rpcSink：写给 Thrift 或者 Avro RPC。

第 2 章

HDFS

本章我们将介绍：
- 读写 HDFS 数据
- 使用 LZO 压缩数据
- 读写序列化文件数据
- 使用 Avro 序列化数据
- 使用 Thrift 序列化数据
- 使用 Protocol Buffers 序列化数据
- 设置 HDFS 备份因子
- 设置 HDFS 块大小

2.1 介绍

Hadoop 分布式文件系统（HDFS）是运行在通用硬件平台上的可容错分布式文件系统。HDFS 优化了大文件的流式读取方式，适用于那些高吞吐并且对延迟性要求相对比较低的场景。此外，HDFS 通过文件一次写入多次读取的简单策略保证了数据的一致性。

HDFS 设计者认为硬件故障是常态，使用了**块复制**的概念，让数据在集群的节点间进行复制。相对于桌面文件系统，HDFS 的数据块会更大些。例如，HDFS 默认的文件块大小是 64 MB。当一个文件存放在 HDFS 上，这个文件会被分割成一个或者多个数据块，并且被分发到集群的节点上。此外将这些数据块备份并分发到集群的节点上也是为了在出现坏硬盘的时候能保证数据的高可用性。每个数据块复制的份数是由**复制因子**决定的。默认的复制因子是 3，意味着每个数据块在集群上会存储 3 份。

最后，Hadoop 将计算放在数据所在的节点上执行，使得应用程序在使用 HDFS 时可以实现高吞吐。换句话说，应用程序可以运行在数据所在的节点机器上而不是传统的将数据复制到应用程序执行的节点上。这个概念称为**数据本地化**。

HDFS 包含以下三个服务。

- NameNode：保存着集群中所有数据块位置的一个目录。
- Secondary NameNode：周期性同步 NameNode 的块索引数据。在同步处理中，Secondary NameNode 下载 NameNode 的 image 文件和 editlogs，并对他们做本地归并，最后再将归并完的 image 文件发回给 NameNode。Secondary NameNode 并不是 NameNode 的热备份。在 NameNode 出故障的时候它并不能工作。
- DataNode：管理着从 NameNode 分配过来的数据块。它并不能感知集群中其他 DataNode 的存在，只和 NameNode 进行交互。

本章将使用文件系统 API，MapReduce，和高级的序列化库有效地写入并存储数据。

Hadoop 0.20.X 不支持文件追加的操作。

2.2 读写 HDFS 数据

HDFS 上的数据读写方式有很多种。我们将使用文件系统的 API 在 HDFS 上创建和写入一个文件，然后再启动一个应用读取 HDFS 上的文件并将其写到本地文件。

准备工作

需要从 `http://www.packtpub.com/support` 下载数据集 `weblog_entries.txt`。

操作步骤

执行以下步骤来读取和写入数据到 HDFS。

1. 下载完测试数据集，我们就可以写一段程序来读取本地文件系统的文件并将其内容写到 HDFS 上。

```
public class HdfsWriter extends Configured implements Tool {
    public int run(String[] args) throws Exception {

        String localInputPath = args[0];
```

```
            Path outputPath = new Path(args[1]);
            Configuration conf = getConf();
            FileSystem fs = FileSystem.get(conf);
            OutputStream os = fs.create(outputPath);
            InputStream is = new BufferedInputStream(
                new FileInputStream(localInputPath));
            IOUtils.copyBytes(is, os, conf);
            return 0;
        }

        public static void main(String[] args) throws Exception {
            int returnCode = ToolRunner.run(
                new HdfsWriter(), args);
            System.exit(returnCode);
        }
    }
```

2. 接下来我们要写一段程序读取刚在 HDFS 上创建的文件并将文件的内容写回本地文件系统。

```
    public class HdfsReader extends Configured implements Tool {

        public int run(String[] args) throws Exception {

            Path inputPath = new Path(args[0]);
            String localOutputPath = args[1];
            Configuration conf = getConf();
            FileSystem fs = FileSystem.get(conf);
            InputStream is = fs.open(inputPath);
            OutputStream os = new BufferedOutputStream(
                 new FileOutputStream(localOutputPath));
            IOUtils.copyBytes(is, os, conf);
            return 0;
        }

        public static void main(String[] args) throws Exception {
            int returnCode = ToolRunner.run(
                new HdfsReader(), args);
            System.exit(returnCode);
        }
    }
```

工作原理

FileSystem 是一个抽象类，代表一个通用文件系统。大多数文件系统访问和操作都可以通过 FileSystem 对象来实现。可以通过 FileSystem.get() 来创建 Hadoop 文件系统的一个实例。The FileSystem.get() 方法会将 Hadoop 配置文件中的 fs.default.name 参数作为 URI，选择初始化一个正确的文件系统。HDFS 的 fs.default.name 参数形式为 hdfs://。

当创建了一个 FileSystem 类的实例后，HdfsWriter 类就会调用 create() 方法在 HDFS 上创建一个文件（如果该文件存在则覆盖）。create() 方法将返回一个可以被普通 Java I/O 方法操作的 OutputStream 对象。同理，HdfsReader 调用方法 open() 读取 HDFS 上的一个文件，并返回一个 InputStream 对象，可以用来读取文件中的内容。

更多参考

FileSystem API 功能很强大。为了证明 API 上其他方法的可用性，我们可以在创建的 HdfsWriter 和 HdfsReader 类上增加一些错误校验。

在文件创建前验证该文件是否已经存在，可以使用：

boolean exists = fs.exists(inputPath);

验证某个路径是否是文件，可以使用：

boolean isFile = fs.isFile(inputPath);

重命名一个已存在的文件，可以使用：

boolean renamed = fs.rename(inputPath, new Path("old_file.txt"));

2.3 使用 LZO 压缩数据

Hadoop 支持许多压缩算法，包含：

- bzip2
- gzip
- DEFLATE

Hadoop 提供了这些算法的 Java 实现，因此我们可以很方便的通过 FileSystem API 对文件进行压缩和解压，也可以通过 MapReduce 输入输出格式化来实现。

但是，使用前面列出的压缩算法对 HDFS 上的数据进行压缩存在着一个问题：这些压缩格式是不可分割的。这意味着，如果一个文件使用上面提到的任何一种编码进行压缩，那么只有整个文件都被读取完才能进行解压。

要理解这个缺点有什么影响，你首先需要知道 Hadoop 是如何决定一个作业所能启动的 mapper 数量的。启动 mapper 的数量约等于输入文件字节数除以数据块的字节数（默认的数据块是 64 MB）。每个 mapper 都将接收到需要处理的数据块，称之为**输入分片**。打个比方，如果一个 MapReduce 的作业输入的是一个没有压缩的 128 MB 文件，那么这个作业会启动 2 个 mapper 来处理（128 MB/64 MB）。

因为使用 bzip2、gzip 和 DEFLATE 算法进行压缩的文件不能分片，所以会把整个文件当成一个分片分配给 mapper。按照前面的例子，如果输入的文件是一个 gzip 压缩过的 128 MB 文件，那么 MapReduce 框架只会启动一个 mapper 来处理。

现在，LZO 如何适用于这些场景？LZO 算法设计的压缩速度和 DEFLATE 差不多，但是 LZO 的解压速度更快。另外，多亏了 Hadoop 成员的辛勤工作，LZO 的压缩文件是可分片的。

 bzip2 在 hadoop 0.21.0 版本上是可分片的，但是存在着一些性能限制。在应用到生产环境前需要做深入的调研。①

准备工作

你可以从 https://github.com/kevinweil/hadoop-lzo 下载 LZO 算法的 Hadoop 实现包。

操作步骤

执行以下步骤，配置 LZO 并对一个文本文件做压缩并建索引。

1. 首先安装 `lzo` 和 `lzo-devel` 包。

在 Red Hat Linux 上使用：

`# yum install liblzo-devel`

在 Ubuntu 上使用：

`# apt-get install liblzo2-devel`

2. 进入 `hadoop-lzo` 源码解压的路径编译这个工程：

```
# cd kevinweil-hadoop-lzo-6bb1b7f/
# export JAVA_HOME=/path/to/jdk/  # ./setup.sh
```

3. 如果编译成功，你将会看到：

`BUILD SUCCESSFUL`

4. 复制编译好的 JAR 文件到集群上的 hadoop 类库目录下。

`# cp build/hadoop-lzo*.jar /path/to/hadoop/lib/`

5. 复制本地类库到集群的 Hadoop 本地类库路径下。

`# tar -cBf - -C build/hadoop-lzo-0.4.15/lib/native/ . | tar -xBvf - -C`

---

① Bzip2 的压缩比更高，但是解压速度慢，译者所在公司大量的使用了 bzip2 的压缩，节约了比较大的存储空间。

/path/to/hadoop/lib/native

6. 接下来，配置 `core-site.xml` 使用 LZO 编码类。

```
<property>
<name>io.compression.codecs</name>
<value>org.apache.hadoop.io.compress.GzipCodec,
        org.apache.hadoop.io.compress.DefaultCodec,
org.apache.hadoop.io.compress.BZip2Codec,
com.hadoop.compression.lzo.LzoCodec,
com.hadoop.compression.lzo.LzopCodec
  </value>
</property>
<property>
  <name>io.compression.codec.lzo.class</name>
  <value>com.hadoop.compression.lzo.LzoCodec</value>
</property>
```

7. 最后，修改 `hadoop-env.sh` 脚本中的下面这些变量：

```
export HADOOP_CLASSPATH=/path/to/hadoop/lib/hadoop-lzo-X.X.XX.jar
export JAVA_LIBRARY_PATH=/path/to/hadoop/lib/native/hadoop-lzo-native-lib:/path/to/hadoop/lib/native/other-native-libs
```

现在测试 LZO 库是否安装正确。

8. 压缩测试数据集：

```
$ lzop weblog_entries.txt
```

9. 将压缩过的 `weblog_entries.txt.lzo` 文件上传到 HDFS：

```
$ hadoop fs -put weblog_entries.txt.lzo /test/weblog_entries.txt.lzo
```

10. 为 `weblog_entries.txt.lzo` 文件创建索引：

```
$ hadoop jar /usr/lib/hadoop/lib/hadoop-lzo-0.4.15.jar com.hadoop.compression.lzo.DistributedLzoIndexer /test/weblog_entries.txt.lzo
```

你将会看到 `/test` 目录下有两个文件

```
$ hadoop fs -ls /test
$ /test/weblog_entries.txt.lzo
$ /test/weblog_entries.txt.lzo.index
```

工作原理

本节涉及的步骤较多。复制完 LZO JAR 文件和本地类库到相应目录，再修改 `core-site.xml` 配置文件中的 `io.compression.codecs` 属性。HDFS 和 MapReduce 都会使用这个配置文件，`io.compression.codecs` 这个配置将决定使用哪种压缩算法。

最后，我们运行了 `DistributedLzoIndexer`。这是一个 MapReduce 应用程序，它会读取一个或者多个 LZO 压缩文件并标记出每个 LZO 文件的块边界。当这个程序作用在 LZO 文件上，只要使用自带的 `LzoTextInputFormat` 输入格式化就可以将 LZO 文件分割给多个 map 处理。

更多参考

除了 `DistributedLzoIndexer`，Hadoop 的 LZO 库也包含一个名为 `LzoIndexer` 的类。`LzoIndexer` 类运行了一个独立的应用程序为 HDFS 的 LZO 文件建立索引。要对 HDFS 上的 `weblog_entries.txt.lzo` 文件建立索引，需运行以下的命令：

```
$ hadoop jar /usr/lib/hadoop/lib/hadoop-lzo-0.4.15.jar com.hadoop.compression.lzo.LzoIndexer /test/weblog_entries.txt.lzo
```

延伸阅读

- 使用 Thrift 序列化数据（2.6 节）。
- 使用 Protocol Buffers 序列化数据（2.7 节）。

2.4 读写序列化文件数据

序列化文件格式是一种灵活的格式，Hadoop 默认支持序列化文件。它能够存储文本和二进制数据。序列化文件以二元键值存储着数据。二元组会被分组到数据块中。这种格式支持记录压缩和块压缩。序列化格式文件是可以被分片的，即使用不可分片的 gzip 格式压缩也是可以被分片的。序列化文件之所以可以被分割，是因为按照记录或者块进行压缩，而不是整个文件压缩。

本节将演示如何写入和读取序列化文件。

准备工作

你需要在 Packt 网站的 http://www.packtpub.com/support 下载数据集 weblog_entries.txt。当然这个文件需要在 HDFS 上可用。你可以通过使用下面的命令将文件上传到 HDFS：

```
$ hadoop fs  put /path/on/local/filesystem/weblog_entries.txt /path/in/hdfs
```

操作步骤

1. 下载完测试数据集，我们就可以写一个程序从 HDFS 上读取一个简单文本文件，并将内容写到一个 HDFS 上的序列化文件里。

```
public class SequenceWriter extends Configured implements Tool {
    public int run(String[] args) throws Exception {
```

```java
        Path inputPath = new Path(args[0]);
        Path outputPath = new Path(args[1]);

        Configuration conf = getConf();
        Job weblogJob = new Job(conf);
        weblogJob.setJobName("Sequence File Writer");
        weblogJob.setJarByClass(getClass());
        weblogJob.setNumReduceTasks(0);
        weblogJob.setMapperClass(IdentityMapper.class);
        weblogJob.setMapOutputKeyClass(LongWritable.class);
        weblogJob.setMapOutputValueClass(Text.class);
        weblogJob.setOutputKeyClass(LongWritable.class);
        weblogJob.setOutputValueClass(Text.class);
        weblogJob.setInputFormatClass(TextInputFormat.class);
        weblogJob.setOutputFormatClass(SequenceFileOutputFormat.class);

        FileInputFormat.setInputPaths(weblogJob, inputPath);
        SequenceFileOutputFormat.setOutputPath(weblogJob, outputPath);

        if(weblogJob.waitForCompletion(true)) {
            return 0;
        }
        return 1;
    }

    public static void main(String[] args) throws Exception {
        int returnCode = ToolRunner.run(new SequenceWriter(), args);
        System.exit(returnCode);
    }
}
```

2. 现在，可以使用 MapReduce 程序读取 HDFS 上的序列化文件，并将其转回普通文本：

```java
public class SequenceWriter extends Configured implements Tool {

    public int run(String[] args) throws Exception {

        Path inputPath = new Path(args[0]);
        Path outputPath = new Path(args[1]);

        Configuration conf = getConf();
        Job weblogJob = new Job(conf);
        weblogJob.setJobName("Sequence File Writer");
        weblogJob.setJarByClass(getClass());
        weblogJob.setNumReduceTasks(0);
        weblogJob.setMapperClass(IdentityMapper.class);
        weblogJob.setMapOutputKeyClass(LongWritable.class);
        weblogJob.setMapOutputValueClass(Text.class);
        weblogJob.setOutputKeyClass(LongWritable.class);
        weblogJob.setOutputValueClass(Text.class);
        weblogJob.setInputFormatClass(SequenceFileInputFormat.class);
        weblogJob.setOutputFormatClass(TextOutputFormat.class);

        SequenceFileInputFormat.addInputPath(weblogJob, inputPath);
        FileOutputFormat.setOutputPath(weblogJob, outputPath);
```

```
            if(weblogJob.waitForCompletion(true)) {
                return 0;
            }
            return 1;
        }

        public static void main(String[] args) throws Exception {
            int returnCode = ToolRunner.run(new SequenceWriter(), args);
            System.exit(returnCode);
        }
    }
```

工作原理

MapReduce 是 HDFS 上转换数据的一种有效方式。这两个 MapReduce 作业编码上非常容易实现，且能够利用集群的分布式能力转换数据。

首先这两个作业都是只有 map 的作业，这意味着 Hadoop 只启动 mapper 来处理测试数据。这是通过设置 reducer 数量为 0 来实现的，可以通过下面这行代码来操作：

`weblogJob.setNumReduceTasks(0);`

接下来，读取序列化文件的作业将文本作为作业的输入，并输出序列化文件。为此，需要将 SequenceWriter 的输入格式化设置为 TextInputFormat.class 类，如下面的代码所示：

`weblogJob.setInputFormatClass(TextInputFormat.class);`

我们还需将输出格式设置为 SequenceFileInputFormat.class 类，如下面的代码所示：

`weblogJob.setOutputFormatClass(SequenceFileOutputFormat.class);`

接下来的程序，我们是要读取序列化文件并且输出普通文本文件。为此，我们要转换输入和输出格式。在读序列化文件的作业中，设置输入格式为 SequenceFileInputFormat.class。

`weblogJob.setInputFormatClass(SequenceFileInputFormat.class);`

设置输出格式为 TextOutputFormat.class。

`weblogJob.setOutputFormatClass(TextOutputFormat.class);`

更多参考

序列化文件有以下三个压缩选项。

- 不压缩：键值对不压缩存储。
- 记录压缩：在 mapper 或 reducer 输出数据的时候压缩。
- 块压缩：整个键值数据块压缩。

你可以在作业初始化的时候使用下面的方法来压缩序列化文件：

```
SequenceFileOutputFormat.setOutputCompression(job, true);
```

接下来，设置要使用的压缩类型，下面的代码设置为记录压缩：

```
SequenceFileOutputFormat.setOutputCompressionType(weblogJob,
SequenceFile.CompressionType.RECORD);
```

或者设置使用块压缩：

```
SequenceFileOutputFormat.setOutputCompressionType(weblogJob,
SequenceFile.CompressionType.BLOCK);
```

最后设置一种压缩编码，比如 gzip 编码：

```
SequenceFileOutputFormat.setOutputCompressorClass(weblogJob,
GzipCodec.class);
```

延伸阅读

在接下来的章节中，我们将继续探索其他的数据序列化库和格式。

- 使用 Avro 序列化数据（2.5 节）。
- 使用 Thrift 序列化数据（2.6 节）。
- 使用 Protocol Buffers 序列化数据（2.7 节）。

2.5 使用 Avro 序列化数据

Avro 被 Apache Avro 网站描述为一个数据序列化系统。Apache Avro 支持与语言无关的文件格式化，包含序列化和 RPC 机制。Avro 有一个灵巧的特性，就是在使用序列化框架的时无需编译任何接口或协议定义文件。

在本节我们将使用 Avro 序列化 Java 对象，并使用 MapReduce 将其写成 HDFS 的文件。

准备工作

需要下载、编译并安装下面的程序。

- 从 http://avro.apache.org/ 上获取 1.5.4 版本的 avro 和 avro-mapred JAR 文件。
- 从 http://www.packtpub.com/support 上获取测试数据文件 weblog_entries.txt。

操作步骤

1. 以下是一个 Java 对象实体，表示 weblog_entries.txt 数据集的一行：

```java
public class WeblogRecord {
    private String cookie;
    private String page;
    private Date date;
    private String ip;

    public WeblogRecord() {

    }
    public WeblogRecord(String cookie, String page, Date date, String ip) {
        this.cookie = cookie;
        this.page = page;
        this.date = date;
        this.ip = ip;
    }
  //getters and setters

    @Override
    public String toString() {
        return cookie + "\t" + page + "\t" + date.toString() + "\t" + ip;
    }

}
```

2. 这将启动一个只有 map 的作业,与生成和读取序列化文件启动的作业一样。但是,我们要写一个 mapper 读取 weblog_entries.txt 取代 IdentityMapper,另外还需创建一个 WeblogRecord 实例。

```java
public class WeblogMapper extends MapReduceBase implements
Mapper<LongWritable, Text, AvroWrapper, NullWritable> {

    private AvroWrapper<WeblogRecord> outputRecord = new AvroWrapper<
WeblogRecord>();

    private WeblogRecord weblogRecord = new WeblogRecord();

    SimpleDateFormat dateFormatter = new SimpleDateFormat("yyyy-MM-dd:HH:
mm:ss");
    public void map(LongWritable key, Text value, OutputCollector<AvroWrapper,
NullWritable> oc, Reporter rprtr)
throws IOException {
        String[] tokens = value.toString().split("\t");
        String cookie = tokens[0];
        String page = tokens[1];
        String date = tokens[2];
        String time = tokens[3];
        String formattedDate = date + ":" + time;
        Date timestamp = null;
        try {
            timestamp = dateFormatter.parse(formattedDate);
        } catch(ParseException ex) {
            // ignore records with invalid dates
```

```
            return;
        }
        String ip = tokens[4];

        weblogRecord.setCookie(cookie);
        weblogRecord.setDate(timestamp);
        weblogRecord.setIp(ip);
        weblogRecord.setPage(page);
        outputRecord.datum(weblogRecord);
        oc.collect(outputRecord, NullWritable.get());
    }

}
```

3. 现在，使用 MapReduce 读取一个文本文件，并序列化和持久化 WeblogRecord 对象：

```
public class AvroWriter extends Configured implements Tool {

    public int run(String[] args) throws Exception {

        Path inputPath = new Path(args[0]);
        Path outputPath = new Path(args[1]);

        Schema schema = ReflectData.get().getSchema(WeblogRecord.class);

        Configuration conf = getConf();
        JobConf weblogJob = new JobConf(conf, getClass());
        weblogJob.setJobName("Avro Writer");
        weblogJob.setNumReduceTasks(0);
        weblogJob.setMapperClass(WeblogMapper.class);
        weblogJob.setMapOutputKeyClass(AvroWrapper.class);
        weblogJob.setMapOutputValueClass(NullWritable.class);
        weblogJob.setInputFormat(TextInputFormat.class);
        AvroJob.setOutputSchema(weblogJob, schema);
        FileInputFormat.setInputPaths(weblogJob, inputPath);
        FileOutputFormat.setOutputPath(weblogJob, outputPath);

        RunningJob job = JobClient.runJob(weblogJob);
        if(job.isSuccessful()) {
            return 0;
        }
        return 1;
    }
    public static void main(String[] args) throws Exception {
        int returnCode = ToolRunner.run(new AvroWriter(), args);
        System.exit(returnCode);
    }
}
```

工作原理

这个 MapReduce 作业读取一个普通文本文件，并将 WeblogRecord 类序列化到一个 Avro 文件。第一步是配置 MapReduce 作业去读取一个普通文本文件，输出格式配置为 Avro 文件格式。

设置输入格式读取一个文本文件：

```
weblogJob.setInputFormat(TextInputFormat.class)
```

编译一个基于 `WeblogRecord` 类的模式,并设置为输出模式:

```
Schema schema = ReflectData.get().getSchema(WeblogRecord.class);
AvroJob.setOutputSchema(weblogJob, schema);
```

接下来,我们使用 Hadoop MapReduce API 实现一个 mapper 并且使用 `AvroWrapper` 类输出 `WeblogRecord` 对象。

`WeblogMapper` 类发送出去的成员对象有:

```
private AvroWrapper<WeblogRecord> outputRecord = new
AvroWrapper<WeblogRecord>();
private WeblogRecord weblogRecord = new WeblogRecord();
```

`WeblogMapper` 的 `map()` 方法发送出去的数据有:

```
outputRecord.datum(weblogRecord);
oc.collect(outputRecord, NullWritable.get());
```

这个只有 map 的作业输出的数据将以 Avro 的文件格式存储。

更多参考

为了读取 `AvroWriter` 作业创建的 Avro 文件,我们只需要更改输入格式化和 mapper 类。首先,设置输入格式化和输入模式。

```
JobConf weblogJob = new JobConf(conf, getClass());
Schema schema = ReflectData.get().getSchema(WeblogRecord.class);
AvroJob.setReflect(weblogJob);
```

接着,按照下面的定义创建一个 mapper 类:

```
public class WeblogMapperAvro extends MapReduceBase
        implements Mapper<AvroWrapper<WeblogRecord>, NullWritable, Text,
NullWritable>
{
public void map(AvroWrapper<WeblogRecord> key, NullWritable value,
OutputCollector<Text, NullWritable> oc, Reporter rprtr) throws
IOException {
        WeblogRecord weblogRecord = key.datum();
        //process the web log record
    }
}
```

延伸阅读

接下来的章节我们将介绍可以在 Hadoop 中使用的其他数据序列化库。

- ❑ 使用 Thrift 序列化数据(2.6 节)。
- ❑ 使用 Protocol Buffers 序列化数据(2.7 节)。

2.6 使用 Thrift 序列化数据

Thrift 是一种跨语言序列化和 RPC 服务框架。Thrift 使用接口定义文件生成包含 Java 在内的许多语言绑定。

本节将示范 Thrift 中的接口定义，生成相应的 Java 绑定，并且通过 MapReduce 使用这些绑定序列化一个 Java 对象到 HDFS 上。

准备工作

需要下载、编译并安装以下内容。

- Hadoop LZO 库。
- 从 `http://thrift.apache.org/` 获取 Apache Thrift 0.70.0 的版本。
- 从 `https://github.com/kevinweil/elephant-bird` 获取最新版本的 Elephant Bird。
- 从 `http://www.packtpub.com/support` 获取测试数据文件 `weblog_entries.txt`。

为了编译安装 Thrift，首先使用 Yum 确认所需的依赖环境已经就绪：

```
# yum install automake libtool flex bison pkgconfig gcc-c++ boost-devel
libevent-devel zlib-devel python-devel ruby-devel openssl-devel
```

接着，编译 Elephant Bird。

```
$ cd /path/to/elephant-bird
$ ant
```

复制 `elephant-bird-X.X.X.jar` 到你开发环境的 classpath 中。

操作步骤

1. 设置目录结构：

```
$ mkdir test-thrift
$ mkdir test-thrift/src
$ mkdir test-thrift/src/thrift
$ mkdir test-thrift/src/java
$ cd test-thrift/src/thrift
```

2. 接着创建一个接口定义：

```
namespace java com.packt.hadoop.hdfs.ch2.thrift

struct WeblogRecord {
  1: optional string cookie,
  2: string page,
```

```
    3: i64 timestamp,
    4: string ip
}
```

保存这个文件到 `test-thrift/src/thrift/` 路径下,并重命名为 `weblog_record.thrift`。

3. 编译生成 `.java` 文件:

```
# thrift --gen java -o src/java/ src/thrift/weblog_record.thrift
```

Thrift 应该会在 `src/java/folder` 生成一个名为 `WeblogRecord.java` 的文件。

4. 现在,我们要写一个 MapReduce 程序读取 HDFS 上的 `weblog_entries.txt`,并使用 Elephant-Bird 的 `LzoThriftBlockOutputFormat` 类序列化 `WeblogRecord` 对象,最终输出成一个 LZO 压缩文件:

```java
public class ThriftMapper extends Mapper<Object, Text, NullWritable,
ThriftWritable<WeblogRecord>> {

    private ThriftWritable<WeblogRecord> thriftRecord =ThriftWritable.
newInstance(WeblogRecord.class);
    private WeblogRecord record = new WeblogRecord();
    private SimpleDateFormat dateFormatter = new SimpleDateFormat("yyyy-
MM-dd:HH:mm:ss");

    @Override
    protected void map(Object key, Text value, Context context) throws
IOException, InterruptedException {
        String[] tokens = value.toString().split("\t");
        String cookie = tokens[0];
        String page = tokens[1];
        String date = tokens[2];
        String time = tokens[3];
        String formatedDate = date + ":" + time;
        Date timestamp = null;
        try {
            timestamp = dateFormatter.parse(formatedDate);
        } catch(ParseException ex) {
            return;
        }
        String ip = tokens[4];
        record.setCookie(cookie);
        record.setPage(page);
        record.setTimestamp(timestamp.getTime());
        record.setIp(ip);
        thriftRecord.set(record);
        context.write(NullWritable.get(), thriftRecord);
    }
}
```

5. 最后我们需要配置 MapReduce 作业。

```java
public class ThriftWriter extends Configured implements Tool {

    public int run(String[] args) throws Exception {
        Path inputPath = new Path(args[0]);
        Path outputPath = new Path(args[1]);

        Configuration conf = getConf();
        Job weblogJob = new Job(conf);
        weblogJob.setJobName("ThriftWriter");
        weblogJob.setJarByClass(getClass());
        weblogJob.setNumReduceTasks(0);
        weblogJob.setMapperClass(ThriftMapper.class);
        weblogJob.setMapOutputKeyClass(LongWritable.class);
        weblogJob.setMapOutputValueClass(Text.class);
        weblogJob.setOutputKeyClass(LongWritable.class);
        weblogJob.setOutputValueClass(Text.class);
        weblogJob.setInputFormatClass(TextInputFormat.class);
        weblogJob.setOutputFormatClass(LzoThriftBlockOutputFormat.class);

        FileInputFormat.setInputPaths(weblogJob, inputPath);
        LzoThriftBlockOutputFormat.setClassConf(
WeblogRecord.class, weblogJob.getConfiguration());
        LzoThriftBlockOutputFormat.setOutputPath(weblogJob, outputPath);

        if(weblogJob.waitForCompletion(true)) {
            return 0;
        }
        return 1;
    }

    public static void main( String[] args ) throws Exception {
        int returnCode = ToolRunner.run( new ThriftWriter(), args);
        System.exit(returnCode);
    }
}
```

工作原理

我们首先要定义和编译一个 Thrift 接口定义文件。这个文件可以用于生成任何 Thrift 支持的语言绑定。

接下来，我们通过 Elephant Bird 构建一个由 Thrift 生成的 `MapReduce` 应用来序列化 WeblogRecord 对象。我们设置 MapReduce 作业的输入格式化，来读取一个简单文本：

```java
weblogJob.setInputFormatClass(TextInputFormat.class);
```

输出格式设置为使用 Thrift 块格式化并且使用 LZO 压缩存储输出数据。

```
LzoThriftBlockOutputFormat.setClassConf(
WeblogRecord.class, weblogJob.getConfiguration());
        LzoThriftBlockOutputFormat.setOutputPath(weblogJob, outputPath);
```

在 mapper 中，我们使用 Elephant Bird 的 `ThriftWritable` 类来绑定 `WeblogRecord` 对象。`ThriftWritable` 类是由 Hadoop 的 `WritableComparable` 类派生而来的，它必须实现所有 MapReduce 产生的键值。我们使用 Thrift 生成的任何类型的绑定，通过 `ThriftWritable` 就可以不用每次都编写普通 `WritableComparable` 类。

在 mapper 中我们初始化了 `ThriftWritable` 和 `WeblogRecord` 实例：

```
private ThriftWritable<WeblogRecord> thriftRecord =
  ThriftWritable.newInstance(WeblogRecord.class);
private WeblogRecord record = new WeblogRecord();
```

然后，我们将 `WeblogRecord` 实例作为参数调用 `thriftRecord` 对象的设置方法。最后 mapper 输出的 `thriftRecord` 对象包含了 `WeblogRecord` 实例。

```
thriftRecord.set(record);
context.write(NullWritable.get(), thriftRecord);
```

延伸阅读

在下面一节中我们将介绍由 Google 开发的一种更流行的序列化框架。

- 使用 Protocol Buffers 序列化数据（2.7 节）。

2.7 使用 Protocol Buffers 序列化数据

Protocol Buffers 是一种跨语言的数据格式化框架。Protocol Buffers 使用接口定义文件生成包含 Java 的许多语言绑定。

本节将示范如何定义一个 Protocol Buffers 消息，生成相应的 Java 绑定，并且通过 MapReduce 使用这些绑定序列化一个 Java 对象到 HDFS 上。

准备工作

需要下载、编译并安装以下内容。

- Hadoop Lzo 库。

- 从 `http://code.google.com/p/protobuf/` 获取 Google Protocol Buffers 2.3.0 的版本。

- Elephant Bird（请看上一节）。

- 从 http://www.packtpub.com/support 获取测试数据文件 weblog_entries.txt。

 注意，你需要安装一个 GNU C/C++编译器来编译 Protocol Buffers 的源码。我们将为 Protocol Buffers 编译源代码。

可以使用 Yum 安装 GNU C/C++编译器，使用 root 用户执行以下脚本完成安装：

```
# yum install gcc gcc-c++ autoconf automake
```

执行以下代码行进行对 Protocol Buffers 的编译和安装：

```
$ cd /path/to/protobuf
$ ./configure
$ make
$ make check
# make install
# ldconfig
```

操作步骤

1. 设置目录结构：

```
$ mkdir test-protobufs
$ mkdir test-protobufs/src
$ mkdir test-protobufs/src/proto
$ mkdir test-protobufs/src/java
$ cd test-protobufs/src/proto
```

2. 接着创建一个协议格式：

```
package example;

option java_package = "com.packt.hadoop.hdfs.ch2";
option java_outer_classname = "WeblogRecord";

message Record {
  optional string cookie = 1;
  required string page = 2;
  required int64 timestamp = 3;
  required string ip = 4;
}
```

把这个文件保存到 test-protobufs/src/proto/ 路径下，并重命名为 weblog_record.proto。

3. 编译 test-protobufs 文件夹下的协议格式。protoc 将在 src/java/ 生成 WeblogRecord.java 文件：

```
$ cd ../../
$ protoc --proto_path=src/proto/ --java_out=src/java/ src/proto/weblog_record.proto
```

4. 现在,我们要写一个 MapReduce 程序读取 HDFS 上的 `weblog_entries.txt`,并使用 Elephant-Bird 的 `LzoProtobufBlockOutputFormat` 类序列化 `WeblogRecord` 对象,最终输出一个 LZO 压缩文件:

```java
public class ProtobufMapper extends Mapper<Object, Text, 
NullWritable, ProtobufWritable<WeblogRecord.Record>> {
    
    private ProtobufWritable<WeblogRecord.Record> protobufRecord = 
ProtobufWritable.newInstance(WeblogRecord.Record.class);
    private SimpleDateFormat dateFormatter = new SimpleDateFormat("yyyy-MM-dd:HH:mm:ss");

    @Override
    protected void map(Object key, Text value, Context context)
    throws IOException, InterruptedException {
        String[] tokens = value.toString().split("\t");
        String cookie = tokens[0];
        String page = tokens[1];
        String date = tokens[2];
        String time = tokens[3];
        String formatedDate = date + ":" + time;
        Date timestamp = null;
        try {
            timestamp = dateFormatter.parse(formatedDate);
        } catch(ParseException ex) {
            return;
        }
        String ip = tokens[4];
        protobufRecord.set(WeblogRecord.Record.newBuilder()
                .setCookie(cookie)
                .setPage(page)
                .setTimestamp(timestamp.getTime())
                .setIp(ip)
                .build());
        context.write(NullWritable.get(), protobufRecord);
    }
}
```

5. 最后我们需要配置 MapReduce 作业。

```java
public class ProtobufWriter extends Configured implements Tool {
    
    public int run(String[] args) throws Exception {
        
        Path inputPath = new Path(args[0]);
        Path outputPath = new Path(args[1]);
        
        Configuration conf = getConf();
```

```
        Job weblogJob = new Job(conf);
        weblogJob.setJobName("ProtobufWriter");
        weblogJob.setJarByClass(getClass());
        weblogJob.setNumReduceTasks(0);
        weblogJob.setMapperClass(ProtobufMapper.class);
        weblogJob.setMapOutputKeyClass(LongWritable.class);
        weblogJob.setMapOutputValueClass(Text.class);
        weblogJob.setOutputKeyClass(LongWritable.class);
        weblogJob.setOutputValueClass(Text.class);
        weblogJob.setInputFormatClass(TextInputFormat.class);
        weblogJob.setOutputFormatClass(
LzoProtobufBlockOutputFormat.class);

        FileInputFormat.setInputPaths(weblogJob, inputPath);
        LzoProtobufBlockOutputFormat.setClassConf(WeblogRecord.
Record.class, weblogJob.getConfiguration());
        LzoProtobufBlockOutputFormat.setOutputPath(weblogJob,
outputPath);

        if(weblogJob.waitForCompletion(true)) {
            return 0;
        }
        return 1;
    }
    public static void main( String[] args ) throws Exception {
        int returnCode = ToolRunner.run(new ProtobufWriter(), args);
        System.exit(returnCode);
    }
}
```

工作原理

我们首先是要定义和编译一个 Protocol Buffers 消息文件。这个定义文件可以用于生成任何 Protocol Buffers 编译器支持的语言绑定。关于消息格式有几点需要注意的。

首先，包定义 `package example;` 与 Java 的包不相关，是 *.proto 文件中定义消息的命名空间。其次，`java_package` 申明的选项才是 Java 中的包定义。最后，申明的 `java_outer_classname` 选项是即将用到的输出类名称。在 `java_outer_classname` 类中，`Record` 类将被定义。

接着，我们需要编写一个 MapReduce 应用序列化由 Protocol Buffers 编译器生成的 `WeblogRecord` 对象。在建立的 MapReduce 作业中，我们设置输入格式读取一个普通的文本文件。

```
weblogJob.setInputFormatClass(TextInputFormat.class);
```

然后，设置输出格式，存储由该作业生成的 Protocol Buffers 块格式的记录，并使用 LZO 压缩。

```
LzoProtobufBlockOutputFormat.setClassConf(WeblogRecord.Record.class,
weblogJob.getConfiguration());
          LzoProtobufBlockOutputFormat.setOutputPath(weblogJob, outputPath);
```

在 mapper 中，我们使用 Elephant Bird 的 `ProtobufWritable` 类来绑定 `WeblogRecord Record` 对象。`ProtobufWritable` 类是由 Hadoop 的 `WritableComparable` 类继承而来的，它必须实现所有 MapReduce 产生的键值。我们使用 `protoc` 生成的任何类型的绑定，通过 `ProtobufWritable` 就可以不用每次都编写普通 `WritableComparable` 类。

在 mapper 中我们实例化一个 `ProtobufWritable` 实例：

```
    private ProtobufWritable<WeblogRecord.Record> protobufRecord =
ProtobufWritable.newInstance(WeblogRecord.Record.class);
```

然后，我们将 `WeblogRecord` 实例作为参数调用 `protobufRecord` 对象的设置方法。最后，mapper 输出 `protobufRecord` 对象。

```
protobufRecord.set(WeblogRecord.Record.newBuilder()
            .setCookie(cookie)
            .setPage(page)
            .setTimestamp(timestamp.getTime())
            .setIp(ip)
            .build());
context.write(NullWritable.get(), protobufRecord);
```

2.8 设置 HDFS 备份因子

HDFS 将文件存储为数据块，并将这些数据块分发到整个集群。HDFS 在设计上容错性好且能运行在廉价硬件上，数据块被复制多份来确保数据的高可用性。复制因子可以通过 HDFS 配置文件的属性来设置，可以为整个集群设置全局的复制因子。对 HDFS 上的每个数据块，会有 n-1 个复制块分布在集群。举个例子，如果复制因子设置为 3，会有一个原始块和两个副本。

准备工作

打开 `hdfs-site.xml`。这个文件可以在 Hadoop 安装目录下的 `conf/` 文件夹中找到。

操作步骤

修改或者添加 `hdfs-site.xml` 文件中下面的属性值：

```
<property>
<name>dfs.replication<name>
<value>3<value>
<description>Block Replication<description>
<property>
```

工作原理

hdfs-site.xml 是 HDFS 的配置文件。修改 hdfs-site.xml 配置文件的 dfs.replication 属性值将会改变所有上传到 HDFS 文件的默认复制份数。

更多参考

你还可以通过 HadoopFS shell 控制台改变每个文件的复制因子。

```
$ hadoop fs -setrep -w 3 /my/file
```

或者，你可以更新一个文件夹下所有文件的复制因子。

```
$ hadoop fs -setrep -w 3 -R /my/dir
```

延伸阅读

- 设置 HDFS 块大小（2.9 节）。

2.9 设置 HDFS 块大小

HDFS 是设计来存储和管理大数据的，因此典型的 HDFS 块大小明显要比你看到的传统文件系统块来得大（如我的手提电脑文件系统的块大小是 4 KB）。块大小的设置用来将大文件切割成一个个数据块，再将这些数据块分发到集群上。比如，集群的块大小设置为 64 MB，一个 128 MB 的文件上传到 HDFS 上，HDFS 会将这个文件切分成 2（128 MB/64 MB）个数据块，再将这两块数据分发到集群的数据节点上。

准备工作

打开 hdfs-site.xml。这个文件可以在 Hadoop 安装目录下的 conf/ 文件夹中找到。

操作步骤

设置 hdfs-size.xml 下面属性值：

```
<property>
<name>dfs.block.size<name>
<value>134217728<value>
<description>Block size<description>
<property>
```

工作原理

`hdfs-site.xml`是HDFS的配置文件。修改`hdfs-site.xml`配置文件的`dfs.block.size`属性值就会改变上传到 HDFS 所有文件的默认块大小。在这个例子中,我们设置`dfs.block.size`为128MB。修改这个值并不会对 HDFS 上现存的任何文件的块大小产生影响,只会改变新上传文件的块大小。

第 3 章

抽取和转换数据

本章我们将介绍：
- 使用 MapReduce 将 Apache 日志转换为 TSV 格式
- 使用 Apache Pig 过滤网络服务器日志中的爬虫访问量
- 使用 Apache Pig 根据时间戳对网络服务器日志数据排序
- 使用 Apache Pig 对网络服务器日志进行会话分析
- 通过 Python 扩展 Apache Pig 的功能
- 使用 MapReduce 及二次排序计算页面访问量
- 使用 Hive 和 Python 清洗、转换地理事件数据
- 使用 Python 和 Hadoop Streaming 执行时间序列分析
- 在 MapReduce 中利用 MultipleOutputs 输出多个文件
- 创建用户自定义的 Hadoop Writable 及 InputFormat 读取地理事件数据

3.1 介绍

对大规模数据进行解析、格式化来使其满足商业需求是一件极具挑战性的工作，相关软件和架构必须满足高可扩展性、高可用性以及运行时间的限制。Hadoop 是一套抽取和转换大规模数据的理想框架。Hadoop 提供了一套非常适合大数据处理的高可扩展性、高可靠性的分布式处理框架。本章将展示使用 MapReduce、Apache Pig、Apache Hive 以及 Python 对数据进行抽取转换的方法。

3.2 使用 MapReduce 将 Apache 日志转换为 TSV 格式

对于将数据转化为**制表符分隔值格式（TSV）**，MapReduce 是一个优秀的工具。只要将数据载入 HDFS，整个 Hadoop 集群就可以并行地转换大规模的数据集。本节将展示从 Apache 访问日志抽取相关记录并以制表符分隔值格式在 HDFS 存储这些记录的方法。

准备工作

你需要从 http://www.packtpub.com/support 下载数据集 apache_clf.txt，并将其导入 HDFS。

操作步骤

执行如下步骤，使用 MapReduce 将 Apache 日志转换为 TSV 格式。

1. 创建正则表达式用于解析 Apache 日志格式：

```
private Pattern p = Pattern.compile("^([\\d.]+) (\\S+) (\\S+) \\[([\\w:/]+\\s
[+\\-]\\d{4})\\] \"(\\w+) (.+?) (.+?)\" (\\d+) (\\d+) \"([^\"]+|(.+?))\"
\"([^\"]+|(.+?))\"", Pattern.DOTALL);
```

2. 创建一个 mapper 类读取日志文件。该 mapper 将 IP 地址作为键，而将时间戳、访问页、http 状态、客户端返回字节数和用户代理（user agent）信息作为值输出：

```
public class CLFMapper extends Mapper<Object, Text, Text, Text>{

    private SimpleDateFormat dateFormatter =
            new SimpleDateFormat("dd/MMM/yyyy:HH:mm:ss Z");
    private Pattern p =
            Pattern.compile("^([\\d.]+) (\\S+) (\\S+)"
            + " \\[([\\w:/]+\\s[+\\-]\\d{4})\\] \"(\\w+) (.+?) (.+?)\" "
            + "(\\d+) (\\d+) \"([^\"]+|(.+?))\" \"([^\"]+|(.+?))\"",
            Pattern.DOTALL);

    private Text outputKey = new Text();
    private Text outputValue = new Text();
    @Override
    protected void map(Object key, Text value, Context
      context) throws IOException, InterruptedException {
        String entry = value.toString();
        Matcher m = p.matcher(entry);
        if (!m.matches()) {
            return;
        }
        Date date = null;
```

```
            try {
                date = dateFormatter.parse(m.group(4));
            } catch (ParseException ex) {
                return;
            }
            outputKey.set(m.group(1)); //ip
            StringBuilder b = new StringBuilder();
            b.append(date.getTime()); //timestamp
            b.append('\t');
            b.append(m.group(6)); //page
            b.append('\t');
            b.append(m.group(8)); //http status
            b.append('\t');
            b.append(m.group(9)); //bytes
            b.append('\t');
            b.append(m.group(12)); //useragent
            outputValue.set(b.toString());
            context.write(outputKey, outputValue);
        }
    }
}
```

3. 随后，创建一个只有 map 的作业，用于执行此转换工作：

```
public class ParseWeblogs extends Configured implements Tool {

    public int run(String[] args) throws Exception {

        Path inputPath = new Path(args[0]);
        Path outputPath = new Path(args[1]);

        Configuration conf = getConf();
        Job weblogJob = new Job(conf);
        weblogJob.setJobName("Weblog Transformer");
        weblogJob.setJarByClass(getClass());
        weblogJob.setNumReduceTasks(0);
        weblogJob.setMapperClass(CLFMapper.class);
        weblogJob.setMapOutputKeyClass(Text.class);
        weblogJob.setMapOutputValueClass(Text.class);
        weblogJob.setOutputKeyClass(Text.class);
        weblogJob.setOutputValueClass(Text.class);
        weblogJob.setInputFormatClass(TextInputFormat.class);
        weblogJob.setOutputFormatClass(TextOutputFormat.class);

        FileInputFormat.setInputPaths(weblogJob, inputPath);
        FileOutputFormat.setOutputPath(weblogJob, outputPath);

        if(weblogJob.waitForCompletion(true)) {
          return 0;
        }
        return 1;
    }

    public static void main( String[] args ) throws Exception {
```

```
            int returnCode = ToolRunner.run(new ParseWeblogs(), args);
            System.exit(returnCode);
    }

}
```

4. 最后，执行此 MapReduce 作业：

```
$ hadoop jar myjar.jar com.packt.ch3.etl.ParseWeblogs /user/hadoop/apache_
clf.txt /user/hadoop/apache_clf_tsv
```

工作原理

首先，mapper 负责抽取 Apache 网络日志中我们需要的信息，并以制表符分隔的格式输出提取后的相关字段。

接着，创建了一个只包含 map 的作业，用于将网络服务器日志转换为制表符分隔的格式。mapper 输出的键值对保存至 HDFS 的一个文件。

更多参考

默认情况下，类 `TextOutputFormat` 使用一个制表符作为键值对的分隔。可以通过设置属性 `mapred.textoutputformat.separator` 修改默认的分隔符。例如，为了得到由一个","分隔的 IP 和时间戳，可以运行下面的命令，重新启动一个作业：

```
$ hadoop jar myjar.jar com.packt.ch3.etl.ParseWeblogs -Dmapred.
textoutputformat.separator=',' /user/hadoop/apache_clf.txt /user/hadoop/csv
```

延伸阅读

本节的制表符分隔值输出将会在以下几节使用。

- ❏ 使用 Apache Pig 过滤网络服务器日志中的爬虫访问量（3.3 节）。
- ❏ 使用 Apache Pig 根据时间戳对网络服务器日志数据排序（3.4 节）。
- ❏ 使用 Apache Pig 对网络服务器日志进行会话分析（3.5 节）。
- ❏ 通过 Python 扩展 Apache Pig 的功能（3.6 节）。
- ❏ 使用 MapReduce 及二次排序计算页面访问量（3.7 节）。

3.3 使用 Apache Pig 过滤网络服务器日志中的爬虫访问量

Apache Pig 是一种用来创建 MapReduce 应用的高级语言。本节将使用 Apache Pig 以及在 Pig 中使用用户自定义过滤器函数（UDF），从样例网络服务器日志数据集中去除所有机器浏览流量。**机器浏览流量**是指非人类的浏览网页行为产生的流量,比如**网络爬虫**(spider)。

准备工作

需要下载、编译并安装如下内容。

- 从 http://pig.apache.org/ 下载 0.8.1 或更高版本 Apache Pig。
- 从 http://www.packtpub.com/support 下载测试数据集 apache_tsv.txt 和 useragent_blacklist.txt。

将 apache_tsv.txt 导入 HDFS 上,将 useragent_blacklist.txt 放置在你的当前工作目录中。

操作步骤

执行如下步骤,使用 Apache Pig UDF 过滤机器浏览流量。

1. 首先,编写一个 Pig UDF,该类继承 Pig 提供的 `FilterFunc` 抽象类。这个类使用用户代理信息字符串过滤网络日志数据集中的相关记录。

```java
public class IsUseragentBot extends FilterFunc {

    private Set<String> blacklist = null;

    private void loadBlacklist() throws IOException {
        blacklist = new HashSet<String>();
        BufferedReader in = new BufferedReader(new FileReader("blacklist"));
        String userAgent = null;
        while ((userAgent = in.readLine()) != null) {
            blacklist.add(userAgent);
        }
    }

    @Override
    public Boolean exec(Tuple tuple) throws IOException {
        if (blacklist == null) {
            loadBlacklist();
        }
        if (tuple == null || tuple.size() == 0) {
            return null;
        }

        String ua = (String) tuple.get(0);
        if (blacklist.contains(ua)) {
            return true;
        }
        return false;
    }
}
```

2. 在当前工作路径创建一个 Pig 脚本。Pig 脚本的开始部分，需要告知 MapReduce 框架 `useragent_blacklist.txt` 在 HDFS 中的路径。

```
set mapred.cache.files '/user/hadoop/blacklist.txt#blacklist';
set mapred.create.symlink 'yes';
```

3. 将包含类 `IsUseragentBot` 的 JAR 文件在 Pig 中注册，并编写根据用户代理信息过滤网络日志的 Pig 脚本。

```
register myudfjar.jar;

all_weblogs = LOAD '/user/hadoop/apache_tsv.txt' AS (ip:
chararray, timestamp:long, page:chararray, http_status:int,
payload_size:int, useragent:chararray);
nobots_weblogs = FILTER all_weblogs BY NOT com.packt.ch3.etl.pig.
IsUseragentBot(useragent);
STORE nobots_weblogs INTO '/user/hadoop/nobots_weblogs';
```

为了运行该 Pig 作业，将 `myudfjar.jar` 放置于 Pig 脚本同一文件夹后执行。

```
$ ls
$ myudfjar.jar filter_bot_traffic.pig
$ pig -f filter_bot_traffic.pig
```

工作原理

通过使用用户自定义函数（UDF），Apache Pig 能进行相关功能的扩展。其中一种方式是通过 Apache Pig 发布版本自带的 Java 抽象类和接口。在本节，我们希望去除所有包含爬虫访问的记录。实现这个目标的一种方法是创建自定义的 Pig 过滤器。

类 `IsUseragentBot` 继承并实现了抽象类 `FilterFunc`，该类重载了 `exec(Tuple t)` 方法。Pig 元组表示一组字段的有序列表。其中，字段类型可以是 Pig 支持的所有基本类型或 `null` 类型。运行的时候，Pig 将数据集中的用户代理信息作为参数，调用 `IsUseragentBot` 类提供的 `exec(Tuple t)` 方法。UDF 通过访问元组中第一个字段，抽取出用户代理信息字符串。如果该用户代理信息字符串是爬虫信息，会返回 `true`，反之返回 `false`。

此外，`IsUseragentBot` UDF 读取名为 `blacklist` 文件，并加载其中的内容到一个 `HashSet` 实例中。名为 `blacklist` 的文件是 `blacklist.txt` 的一个符号链接。该文件通过**分布式缓存**机制被分发到集群中的每一个节点。设置一个文件到分布式缓存并设置符号链接，可以通过设置如下 MapReduce 属性完成：

```
set mapred.cache.files '/user/hadoop/blacklist.txt#blacklist';
set mapred.create.symlink 'yes';
```

这里需要注意的是，这些属性不是属于 Pig 的。这些属性是作用于 MapReduce 框架的。

因此，你可以对任何 MapReduce 作业设置这些属性，用于分布式缓存加载文件。

接下来，需要告知 Pig 包含 `IsUseragentBot` UDF 的 JAR 文件位置：

`register myudfjar.jar;`

最后，通过 Java 类名调用 UDF。作业运行时，Pig 将实例化一个 `IsUseragentBot` 类的实例，调用 `exec(Tuple t)` 方法，从 `all_weblogs` 关系中获得相关记录。

更多参考

自 Pig 0.9 版本以后，Pig UDF 不需要设置属性 `mapred.cache.files` 和 `mapred.create.symlink` 就能直接访问分布式缓存。在 Pig 中，大多数用于创建 UDF 的抽象类如今都包含一个 `List<String> getCacheFiles()` 方法，重载该方法能够将 HDFS 中的文件载入至分布式缓存中。例如，类 `IsUseragentBot` 添加如下方法，达到将文件 `blacklist.txt` 加载至分布式缓存的目的：

```
@Override
public List<String> getCacheFiles() {
    List<String> list = new ArrayList<String>();
    list.add("/user/hadoop/blacklist.txt#blacklist");
    return list;
}
```

延伸阅读

Apache Pig 将用于本章以下几节中。

- 使用 Apache Pig 根据时间戳对网络服务器日志数据排序（3.4 节）。
- 使用 Apache Pig 对网络服务器日志进行会话分析（3.5 节）。
- 通过 Python 扩展 Apache Pig 的功能（3.6 节）。
- 使用 MapReduce 及二次排序计算页面访问量（3.7 节）。

3.4 使用 Apache Pig 根据时间戳对网络服务器日志数据排序

对数据进行排序是一种常见的数据变换处理技术。本节将展示利用 Hadoop 集群的分布式处理能力，通过编写一个简单的 Pig 脚本对数据集进行排序。

准备工作

需要下载、编译并安装如下内容。

- 从 `http://pig.apache.org/` 下载 0.8.1 或更高版本 Apache Pig。

- 从 http://www.packtpub.com/support 下载测试数据集 apache_nobots_tsv.txt。

操作步骤

执行如下步骤，使用 Apache Pig 对数据进行排序。

1. 首先，将网络服务器日志数据加载至一个 Pig 关系中：

```
nobots_weblogs = LOAD '/user/hadoop/apache_nobots_tsv.txt' AS (ip: chararray,
timestamp:long, page:chararray, http_status:int, payload_size:int, useragent:
chararray);
```

2. 按照 timestamp 字段对网络服务器日志进行升序排列：

```
ordered_weblogs = ORDER nobots BY timestamp;
```

3. 将排序后的结果存储至 HDFS：

```
STORE ordered_weblogs INTO '/user/hadoop/ordered_weblogs';\
```

4. 运行 Pig 作业：

```
$ pig -f ordered_weblogs.pig
```

工作原理

在一个分布式的环境下，排序数据是一件至关重要的事。Pig 的关系操作 ORDER BY 提供了对整个数据集排序的能力。这意味着出现在输出文件 part-00000 记录的时间戳都小于输出文件 part-00001 记录的时间戳（因为我们的数据是按照时间戳排序）。

更多参考

Pig 的 ORDER BY 关系操作既支持多字段的排序也支持降序排序。例如，根据 ip 以及 timestamp 字段对 nobots 关系进行排序，可以使用如下表达式：

```
ordered_weblogs = ORDER nobots BY ip, timestamp;
```

对 nobots 关系进行时间戳字段的降序，可以使用 desc 选项：

```
ordered_weblogs = ORDER nobots timestamp desc;
```

延伸阅读

Apache Pig 将用于以下几节中。

- 使用 Apache Pig 对网络服务器日志进行会话分析（3.5 节）。
- 通过 Python 扩展 Apache Pig 的功能（3.6 节）。
- 使用 MapReduce 及二次排序计算页面访问量（3.7 节）。

3.5 使用 Apache Pig 对网络服务器日志进行会话分析

一个会话表达一个用户持续地与网络交互的行为，当连接发生超时，用户的会话也将随之结束。当用户离线一段时间后再重新进入网站，新的会话又随之开始。本节将使用 Apache Pig 以及 Pig 的用户自定义函数（UDF）生成文件 `apache_nobots_tsv.txt` 的一个子集记录，用于标记对于特定的 IP 的会话开始信息。

准备工作

需要下载、编译并安装如下内容。

- 从 `http://pig.apache.org/` 下载 0.8.1 或更高版本 Apache Pig。
- 从 `http://www.packtpub.com/support` 下载测试数据集 `apache_nobots_tsv.txt`。

操作步骤

执行如下步骤，使用 Apache Pig 对网络服务器日志数据进行会话信息标记。

1. 创建一个 Pig UDF 只发送一个会话的第一条记录。该 UDF 需要继承 Pig 抽象类 `EvalFunc` 并实现 Pig 接口 `Accumulator`。该类负责对网络服务器日志数据集进行会话逻辑应用分析。

```java
public class Sessionize extends EvalFunc<DataBag> implements
Accumulator <DataBag> {

    private long sessionLength = 0;
    private Long lastSession = null;
    private DataBag sessionBag = null;

    public Sessionize(String seconds) {
        sessionLength = Integer.parseInt(seconds) * 1000;
        sessionBag = BagFactory.getInstance().newDefaultBag();
    }

    @Override
    public DataBag exec(Tuple tuple) throws IOException {
        accumulate(tuple);
        DataBag bag = getValue();
        cleanup();
        return bag;
    }

    @Override
    public void accumulate(Tuple tuple) throws IOException {
        if (tuple == null || tuple.size() == 0) {
            return;
        }
```

```
            DataBag inputBag = (DataBag) tuple.get(0);
            for(Tuple t: inputBag) {
                Long timestamp = (Long)t.get(1);
                if (lastSession == null) {
                    sessionBag.add(t);
                }
                else if ((timestamp - lastSession) >= sessionLength) {
                    sessionBag.add(t);
                }
                lastSession = timestamp;
            }
        }

        @Override
        public DataBag getValue() {
            return sessionBag;
        }   @Override
        public void cleanup() {
            lastSession = null;
            sessionBag = BagFactory.getInstance().newDefaultBag();
        }
}
```

2. 创建 Pig 脚本用于加载网络服务器日志并按照 IP 地址对其进行分组操作：

```
register myjar.jar;
define Sessionize com.packt.ch3.etl.pig.Sessionize('1800'); /* 30 
minutes */

nobots_weblogs = LOAD '/user/hadoop/apache_nobots_tsv.txt' AS
(ip: chararray, timestamp:long, page:chararray, http_status:int,
payload_size:int, useragent:chararray);

ip_groups = GROUP nobots_weblogs BY ip;
```

3. 最后，使用 Pig 表达式对所有与特定 IP 关联的记录按照时间排序。发送排序后的记录至 `Sessionize` UDF：

```
sessions = FOREACH ip_groups {
                ordered_by_timestamp = ORDER nobots_weblogs BY timestamp;
                GENERATE FLATTEN(Sessionize(ordered_by_timestamp));
           }

STORE sessions INTO '/user/jowens/sessions';
```

4. 复制包含类 `Sessionize` 的 JAR 文件至当前工作目录，并运行该 Pig 脚本：

```
$ pig -f sessionize.pig
```

工作原理

首先创建的 UDF 继承了抽象类 `EvalFunc` 并实现了接口 `Accumulator`。类 `EvalFunc`

创建用于 Pig 脚本的自定义的函数。数据将通过 exec(Tuple t) 处理后传至 UDF。Accumulator 的构造函数是自定义 eval 方法的可选项，允许 Pig 优化数据流以及 UDF 的内存使用。与 EvalFunc 类运行方式类似，Accumulator 函数允许发送数据的子集给 UDF 而非发送所有的数据。

接下来，Pig 脚本根据 IP 对所有的网络服务器日志记录进行分组并按照时间戳对其排序。数据需要按照时间戳排序是因为 Sessionize UDF 根据排序后的时间戳序列确定每个会话的开始时间。

随后，通过调用 Sessionize 别名生成所有与特定 IP 关联的会话。

最后，使用 FLATTEN 操作嵌套展开 UDF 输出的 DataBag 中包含的元组。

延伸阅读

❑ 通过 Python 扩展 Apache Pig 的功能（3.6 节）。

3.6 通过 Python 扩展 Apache Pig 的功能

本节将使用 Python 创建一个简单的 Apache Pig 用户自定义函数，用于计算在一个 Pig DataBag 中的记录条数。

准备工作

需要下载、编译并安装如下内容。

❑ 从 http://www.jython.org/ 下载 Jython 2.52。

❑ 从 http://pig.apache.org/ 下载 0.8.1 或更高版本 Apache Pig。

❑ 从 http://www.packtpub.com/support 下载测试数据集 apache_nobots_tsv.txt。

本节需要独立的 Jython JAR 文件。为了创建该文件，下载 Jython 的 Java 安装程序，运行安装程序，并在安装菜单选择 **Standalone**。

```
$ java -jar jython_installer-2.5.2.jar
```

增加 Jython 可独立存在的 JAR 文件到 Apache Pig 的路径下面：
```
$ export PIG_CLASSPATH=$PIG_CLASSPATH:/path/to/jython2.5.2/jython.jar
```

操作步骤

使用 Python 创建一个 Apache Pig UDF 的步骤如下。

1. 首先，创建一个简单的 Python 函数，用于计算在一个 Pig DataBag 中的记录条数：

```
#!/usr/bin/python

@outputSchema("hits:long")
def calculate(inputBag):
  hits = len(inputBag)
  return hits
```

2. 创建 Pig 脚本，根据 IP 以及访问页面对所有的网络服务器日志记录进行分组。发送分组后的网络服务器日志至 Python 函数：

```
register 'count.py' using jython as count;

nobots_weblogs = LOAD '/user/hadoop/apache_nobots_tsv.txt' AS
(ip: chararray, timestamp:long, page:chararray, http_status:int,
payload_size:int, useragent:chararray);

ip_page_groups = GROUP nobots_weblogs BY (ip, page);

ip_page_hits = FOREACH ip_page_groups GENERATE FLATTEN(group),
count.calculate(nobots_weblogs);

STORE ip_page_hits INTO '/user/hadoop/ip_page_hits';
```

工作原理

首先，创建简单的 Python 函数对 Pig DataBag 长度进行计算。此外，Python 脚本包含 Python decorator `@outputSchema("hits:long")`，指导 Pig 如何解释 Python 函数返回的数据。在此情况下，我们希望 Pig 使用该函数存储数据并返回 `hits` 值，其中 `hits` 就好比 Java 中的 `Long` 类型。

随后，Pig 脚本使用如下语句对 Python UDF 进行注册：

```
register 'count.py' using jython as count;
```

最后，在 Pig DataBag 中我们通过别名 `count` 来调用 `calculate()` 方法：

```
count.calculate(nobots_weblogs);
```

3.7 使用 MapReduce 及二次排序计算页面访问量

在典型的 MapReduce 作业中，键值对从 mapper 发出，通过洗牌（shuffle）、排序后，最后传输至 reducer。MapReduce 框架未在处理过程中对 reducer 接收的值进行排序。然而，在某些情况下我们需要对 reducer 接收到的值进行排序，例如计算页面访问量的情况。

为了计算页面访问量，需要计算每个页面独立的 IP 数。其中一种方法是 mapper 输出 page、IP 组成的键值对。然后在 reducer 端，将同一页面关联的 IP 存储在一个集合中。但是该方法缺乏可扩展性的。试想如果日志中存在大量的网页被不同独立 IP 访问，将会发生怎样的情况呢？整个独立 IP 集合将无法在内存存储。

MapReduce 框架提供了解决此类问题的方法。在本节，使用称之为**二次排序**（secondary sort）的方法，MapReduce 应用对 reducer 中的值进行排序。随后，在 reducer 端记下上次读取的 IP 值，判断是否重复，再声明一个计数器用于记录独立 IP 数，避免将所有独立 IP 放在内存中。

准备工作

你需要从 `http://www.packtpub.com/support` 下载测试数据集 `apache_nobots_tsv.txt`，并放置在 HDFS 上。

操作步骤

下面步骤表明如何在 MapReduce 中实现二次排序用于计算页面访问量。

1. 创建一个类实现 Hadoop 提供的接口 `WritableComparable`，存储相应键值以及用于排序的字段：

```java
public class CompositeKey implements WritableComparable {

    private Text first = null;
    private Text second = null;

    public CompositeKey() {

    }
    public CompositeKey(Text first, Text second) {
        this.first = first;
        this.second = second;
    }

    //...getters and setters

    public void write(DataOutput d) throws IOException {
        first.write(d);
        second.write(d);
    }
    public void readFields(DataInput di) throws IOException {
        if (first == null) {
            first = new Text();
        }
        if (second == null) {
            second = new Text();
        }
        first.readFields(di);
```

```java
            second.readFields(di);
        }

        public int compareTo(Object obj) {
            CompositeKey other = (CompositeKey) obj;
            int cmp = first.compareTo(other.getFirst());
            if (cmp != 0) {
                return cmp;
            }
            return second.compareTo(other.getSecond());
        }

        @Override
        public boolean equals(Object obj) {
            CompositeKey other = (CompositeKey)obj;
            return first.equals(other.getFirst());
        }

        @Override
        public int hashCode() {
            return first.hashCode();
        }
    }
```

2. 编写 Mapper 和 Reducer 类。Mapper 使用类 CompositeKey 存储两个字段,第一个字段是 page,对输出的 **mapper** 数据进行分组划分,第二个字段是 ip,在数据到达 reducer 后按照这个字段进行排序。

```java
public class PageViewMapper extends Mapper<Object, Text, CompositeKey, Text> {
    private CompositeKey compositeKey = new CompositeKey();
    private Text first = new Text();
    private Text second = new Text();
    private Text outputValue = new Text();
    @Override
    protected void map(Object key, Text value, Context
      context) throws IOException, InterruptedException {
        String[] tokens = value.toString().split("\t");
        if (tokens.length > 3) {
            String page = tokens[2];
            String ip = tokens[0];
            first.set(page);
            second.set(ip);
            compositeKey.setFirst(first);
            compositeKey.setSecond(second);
            outputValue.set(ip);
            context.write(compositeKey, outputValue);
        }
    }
}

public class PageViewReducer extends Reducer<CompositeKey, Text,
Text, LongWritable> {
    private LongWritable pageViews = new LongWritable();
```

```
            @Override
            protected void reduce(CompositeKey key, Iterable<Text>
              values, Context context)
              throws IOException, InterruptedException {
                String lastIp = null;
                long pages = 0;
                for(Text t : values) {
                    String ip = t.toString();
                    if (lastIp == null) {
                        lastIp = ip;
                        pages++;
                    }
                    else if (!lastIp.equals(ip)) {
                        lastIp = ip;
                        pages++;
                    }
                    else if (lastIp.compareTo(ip) > 0) {
                        throw new IOException("secondary sort failed");
                    }
                }
                pageViews.set(pages);
                context.write(key.getFirst(), pageViews);
            }
        }
```

3. 创建三个类，分别对 mapper 输出的数据划分、分组以及排序。MapReduce 框架将使用这些类。首先，编写根据 page 字段对 mapper 输出的数据进行划分的类：

```
static class CompositeKeyParitioner extends Partitioner<CompositeKey, Writable> {

            @Override
            public int getPartition(CompositeKey key, Writable value, int numParition){
                return (key.getFirst().hashCode() &  0x7FFFFFFF) % numParition;
            }
        }
```

4. 编写比较器，将所有相同的键分组到一块。

```
static class GroupComparator extends WritableComparator {
            public GroupComparator() {
                super(CompositeKey.class, true);
            }

            @Override
            public int compare(WritableComparable a, WritableComparable b) {
                CompositeKey lhs = (CompositeKey)a;
                CompositeKey rhs = (CompositeKey)b;
                return lhs.getFirst().compareTo(rhs.getFirst());
            }
        }
```

5. 编写二次排序比较器，对发送至 reducer 的数据进行排序。

```
static class SortComparator extends WritableComparator {
            public SortComparator() {
```

```
            super(CompositeKey.class, true);
    }
    @Override
    public int compare(WritableComparable a, WritableComparable b) {
        CompositeKey lhs = (CompositeKey)a;
        CompositeKey rhs = (CompositeKey)b;
        int cmp = lhs.getFirst().compareTo(rhs.getFirst());
        if (cmp != 0) {
            return cmp;
        }
        return lhs.getSecond().compareTo(rhs.getSecond());
    }
}
```

6. 最后，编写程序用于启动一个常规的 MapReduce 作业，但需要告知 MapReduce 框架使用自定义的分组器类和比较器类：

```
public int run(String[] args) throws Exception {

    Path inputPath = new Path(args[0]);
    Path outputPath = new Path(args[1]);

    Configuration conf = getConf();
    Job weblogJob = new Job(conf);
    weblogJob.setJobName("PageViews");
    weblogJob.setJarByClass(getClass());
    weblogJob.setMapperClass(PageViewMapper.class);
    weblogJob.setMapOutputKeyClass(CompositeKey.class);
    weblogJob.setMapOutputValueClass(Text.class);

    weblogJob.setPartitionerClass(CompositeKeyParitioner.class);
    weblogJob.setGroupingComparatorClass(GroupComparator.class);
    weblogJob.setSortComparatorClass(SortComparator.class);

    weblogJob.setReducerClass(PageViewReducer.class);
    weblogJob.setOutputKeyClass(Text.class);
    weblogJob.setOutputValueClass(Text.class);
    weblogJob.setInputFormatClass(TextInputFormat.class);
    weblogJob.setOutputFormatClass(TextOutputFormat.class);

    FileInputFormat.setInputPaths(weblogJob, inputPath);
    FileOutputFormat.setOutputPath(weblogJob, outputPath);

    if(weblogJob.waitForCompletion(true)) {
        return 0;
    }
    return 1;
}
```

工作原理

首先，创建的类 CompositeKey 实现了 Hadoop 提供的 WritableComparable 接口，以便

我们像其他通常的 Hadoop `WritableComparable` 接口（如 `Text` 及 `IntWritable`）一样使用 `CompositeKey` 类。`CompositeKey` 类包含两个 `Text` 对象：第一个 `Text` 对象用来划分、分组 `mapper` 发送的键值对，第二个 `Text` 对象用于完成二次排序。

随后，`mapper` 类发送的键值对以 `CompositeKey`（其中包含 `page`、IP 字段）作为键，IP 作为值。之后，`reducer` 类接收一个 `CompositeKey` 对象以及排序后的 IP 列表。如果当前出现的 IP 和之前出现的 IP 不同，对应的独立 IP 计数器将会增加。

在创建 `mapper` 和 `reducer` 类之后，又编写了划分、分组和排序数据的三个类。类 `CompositeKeyPartitioner` 负责划分 `mapper` 发出的数据。本节，需要同样的访问页面位于同一个分组。因此只需要根据 `CompositeKey` 类的第一个字段进行划分位置的计算。

创建的类 `GroupComparator` 与类 `CompositeKeyPartitioner` 具有相同的逻辑。我们希望划分在同一组的网页键值被同一个 `reducer` 处理。因此，分组比较器只使用类 `CompositeKey` 的第一个成员变量作比较。

最后，类 `SortComparator` 负责对所有发送至 `reducer` 的值进行排序。正如你所看到的方法 `compare(WritableComparable a, WritableComparable b)` 所示，只接收那些需要发往每个 `reducer` 的 key。这就是为什么 `mapper` 输出的每一条记录的键都要包含 IP 的原因。`SortComparator` 先比较类 `CompositeKey` 的第一个字段，再比较第二个字段，从而保证每个 `reducer` 接收到的值是有序的。

延伸阅读

- 创建用户自定义的 Hadoop Writable 及 InputFormat 读取地理事件数据（3.11 节）。

3.8 使用 Hive 和 Python 清洗、转换地理事件数据

本节通过用户自定义 Python 脚本，在 Hive 中使用一些操作输入、输出数据。该脚本对每一行数据进行一些简单的删减操作，并将稍微修改后的行记录版本存入 Hive 表中。

准备工作

需要下载、编译并安装如下内容。

- 从 `http://hive.apache.org/` 下载 0.7.1 以上版本的 Apache Pig。
- 从 `http://www.packtpub.com/support` 下载测试数据集 `Nigeria_ACLED.csv` 和 `Nigeria_ACLED_cleaned.tsv`。

❑ Python 2.7 以上版本。

本节需要将 `Nigeria_ACLED.csv` 文件加载至 Hive 表 `acled_nigeria` 中，且相关字段对应如下数据类型。

在 Hive 客户端输入如下命令：

```
describe acled_nigeria
```

将会返回如下内容：

```
OK
loc   string
event_date   string
year   string
event_type   string
actor   string
latitude   double
longitude  double
source string
fatalities   string
```

操作步骤

通过如下步骤，使用 Python 及 Hive 转换数据。

1. 在当前工作路径创建文件 `clean_and_transform_acled.hql`，添加内联创建转换语法：

```
SET mapred.child.java.opts=-Xmx512M;

DROP TABLE IF EXISTS acled_nigeria_cleaned;
CREATE TABLE acled_nigeria_cleaned (
    loc STRING,
    event_date STRING,
    event_type STRING,
    actor STRING,
    latitude DOUBLE,
    longitude DOUBLE,
    source STRING,
    fatalities INT
) ROW FORMAT DELIMITED;

ADD FILE ./clean_acled_nigeria.py;
INSERT OVERWRITE TABLE acled_nigeria_cleaned
    SELECT TRANSFORM(
            if(loc != "", loc, 'Unknown'),
            event_date,
            year,
            event_type,
            actor,
```

```
            latitude,
            longitude,
            source,
            if(fatalities != "", fatalities, 'ZERO_FLAG'))
    USING 'python clean_acled_nigeria.py'
    AS (loc, event_date, event_type, actor, latitude, longitude, source, fatalities)
    FROM acled_nigeria;
```

2. 在与 clean_and_transform_acled.hql 同样的工作目录下,创建另一个文件 clean_acled_nigeria.py,并添加如下的 Python 代码从标准输入读取数据:

```
#!/usr/bin/env python
import sys

for line in sys.stdin:
    (loc, event_date, year, event_type, actor, lat, lon, src, fatalities) = line.strip().split('\t')
    if loc != 'LOCATION': #remove header row
      if fatalities == 'ZERO_FLAG':
        fatalities = '0'
        print '\t'.join([loc, event_date, event_type, \ actor, lat, lon, src, fatalities]) #strip out year
```

 Python 对于不一致的缩进是敏感的。因此,如果你直接复制粘贴上面的代码,请特别小心。

3. 在系统 shell 中添加 -f 的选项,就可以使用 Hive 客户端运行此脚本:

```
$ hive -f clean_and_transform_acled.hql
```

4. 为了确保脚本的正确执行完成,在 Hive 客户端运行如下带 -e 选项的命令:

```
hive -e "select count(1) from acled_nigeria_cleaned"
```

Hive 将统计共 2931 行记录。

工作原理

首先从我们创建的 Hive 脚本开始。第一行是对执行过程的 JVM 堆大小进行简单的设置。可以为你的集群设置任意合适的值。对于 ACLED Nieria 数据集,512 MB 的内存已经足够大了。

接下来删除表 acled_nigeria_cleaned 并创建同名的表。由于分隔符",", "\n" 都是由 ROW FARMAT 设定的默认字段分隔符、行分隔符,且 ACLED Nieria 数据集也是以该形式表达,因此我们可以忽略字段分隔符","以及行分隔符"\n"。

一旦此表定义好之后,需要定义 SELECT 语句转换并输出其中的数据。一般约定:在这些语句之前需要添加 SELECT 依赖的脚本。语句 ADD FILE ./clean_acled_nigeria.py 告知 Hive 将脚本从本地文件系统加载至分布式缓存供 MapReduce 任务使用。

  `SELECT` 语句使用 Hive 提供的 `TRANSFROM` 操作将列按照制表符分隔，并把每一列都转化为字符串，当遇到 null 则转成 "\n"。其中，列 `loc` 和列 `fatalities` 需要对空串进行条件检查，如果是空串需要设定默认值。

  使用 `USING` 操作符指定加载自定义脚本所需运行的 `TRANSFORM` 操作。在 Hive 中，自定义脚本在使用 `USING` 操作符之前，需要变换相应的列。如果脚本文件分发至分布式缓存，且集群中的每个节点已安装 Python，MapReduce JVM 就能并行地执行相应脚本以及读取行数据。`AS` 操作符包含了与接收数据的 Hive 表 `acled_nigeria_cleaned` 列名一致的字段列表。

  Python 脚本是非常容易理解的。`#!/usr/bin/env python` 是一个提示语句，告知 shell 如何执行此脚本。表中的每行记录是通过标准输入得到的。调用 `strip()` 方法去除一行数据的开头及结尾的空白字符。将记录分隔并存储在已命名的数组中。每一列字段保存为一个已命名变量中。原始的 ACLED Nigeria 作为 Hive 表的输入，包含表头内容，这些内容是需要去除的。为此，首先检查 `LOCATION` 的值是否为 `loc`，表明该行是否为需要忽略的表头部分。

  如果这行通过上面的检测，需要查看 `fatalities` 的值是否为 `ZERO_FLAG`。`ZERO_FLAG` 这个值已经在前面的 Hive 脚本设置了。如果检测到 `fatalities` 的值满足条件，则将其 `fatalities` 的值赋值为字符串 "0"。

  最后，按照输入同样的顺序输出各个字段，但需要排除 `year` 字段。每行的相关内容将保存在 `acled_nigeria_cleaned` 表内。

更多参考

  本节还有更多参考的内容。下面是一些附加内容，一般来说有助于你使用 Hive 的 `TRANSFORM`、`USING`、`AS` 等操作和 ETL。

定义每列数据类型为字符串

  这确实有点违反直观认识，且在任何的 Hive 文档中也无法找寻出处。如果刚开始把 Hive 表中新进数据的每个字段都映射为字符串，将会极大地有助于数据的验证和调试。你可以使用 Hive 的 `STRING` 类型实现将任意输入数据表示为一个简单的脚本或者一个 Hive QL 语句。如果直接准确地映射数据集，那么会因为错误的格式输入，造成处理不灵活。比如预期为数字类型的字段有可能出现异常字符，这类错误会使数据分析变得不容易。将原始数据字段表示为字符串数据类型，允许自定义函数检测非法的数据并决定如何处理这些数据。另外，当处理 CSV 或制表符分隔数据的时候，只要存在一个与 Hive 表结构声明稍微错位的 `INT` 或 `FLOAT` 类型，当数据中包含一个 `STRING`，该行就有可能被映射为 `NULL`。将原始表结构映射为字符串能迅速地显示出列错位的情况。这是一个见仁见智的问题，只适

使用 AS 关键字对值进行类型转换

本节利用标准输出从 Python 脚本输出字符串类型。Hive 将在接收表中尝试将这些输出数据转换为合适的数据类型。不使用 AS 操作符显式地对每个字段进行类型转换的优点是节约了时间以及代码空间,缺点是如果一个值被转换为不合适的类型并不会失败。比如,输出 HI THERE 至一个数字类型字段,将在该字段插入一个 NULL。这将导致随后针对该表的 SELECT 语句会产生不正确的结果。

在本地测试脚本

这是不言自明的。在命令行调试脚本比通过 MapReduce 任务错误日志容易太多。该方法虽然不会避免由于可扩展性或数据正确性带来的问题,但能大幅度减少调试编译的时间以及关于控制流程的问题。

3.9 使用 Python 和 Hadoop Streaming 执行时间序列分析

本节涉及如何使用 Hadoop Streaming 以及 Python 对清洗过的 ACLED Nigeria 数据集进行基本的时序分析。该程序的目的是输出一个有序的时间列表,该列表上的时间为尼日利亚政府收回的各地领土的时间。

在本节中,使用的尼日利亚冲突数据由武装冲突与事件数据集收集团队提供。

准备工作

需要下载、编译并安装如下内容。

- 从 http://hive.apache.org/ 下载 0.7.1 以上版本的 Apache Pig。
- 从 http://www.packtpub.com/support 下载测试数据集 Nigeria_ACLED_cleaned.tsv。
- Python 2.6 以上版本。

操作步骤

执行以下步骤,使用 Python 及 Hadoop Streaming。

1. 创建名为 run_location_regains.sh 的脚本,用于运行 Streaming 作业。修改 streaming JAR 路径使之与 hadoop-streaming.jar 文件绝对路径匹配是十分重要的。文件

hadoop-streaming.jar 的路径是随着 Hadoop 发布的版本而不同的。

```bash
#!/bin/bash
$HADOOP_HOME/bin/hadoop jar $HADOOP_HOME/contrib/streaming/hadoop-streaming-0.20.2-cdh3u1.jar \
    -input /input/acled_cleaned/Nigeria_ACLED_cleaned.tsv \
    -output /output/acled_analytic_out \
    -mapper location_regains_mapper.py \
    -reducer location_regains_by_time.py \
    -file location_regains_mapper.py \
    -jobconf stream.num.map.output.key.fields=2 \
    -jobconf map.output.key.field.separator=\t \
    -jobconf num.key.fields.for.partition=1 \
    -jobconf mapred.reduce.tasks=1
```

2. 创建 Python 文件 location_regains_mapper.py，添加如下内容：

```python
#!/usr/bin/python
import sys

for line in sys.stdin:
    (loc, event_date, event_type, actor, lat, lon, src, fatalities) = line.strip().split('\t');
    (day,month,year) = event_date.split('/')
    if len(day) == 1:
        day = '0' + day
    if len(month) == 1:
        month = '0' + month;
    if len(year) == 2:
        if int(year) > 30 and int(year) < 99:
            year = '19' + year
        else:
            year = '20' + year
    event_date = year + '-' + month + '-' + day
    print '\t'.join([loc, event_date, event_type]);
```

3. 创建 Python 文件 location_regains_by_time.py，添加如下内容：

```python
#!/usr/bin/python
import sys

current_loc = "START_OF_APP"
govt_regains=[]
for line in sys.stdin:
  (loc,event_date,event_type) = line.strip('\n').split('\t')
  if loc != current_loc:
    if current_loc != "START_OF_APP":
      print current_loc + '\t' + '\t'.join(govt_regains)
    current_loc = loc
    govt_regains = []
  if event_type.find('regains') != -1:
     govt_regains.append(event_date)
```

4. 在包含刚才所创建的 Python 脚本的当前工作路径中运行 shell 脚本：

```
./run_location_regains.sh
```

你将会在命令行看到作业开始以及成功完成的信息:
```
INFO streaming.StreamJob: Output: /output/acled_analytic_out
```

工作原理

shell 脚本设置 Hadoop Streaming JAR 的路径,同时传递程序运行需要的一些参数。下面的表对每个参数进行解释。

参　　　数	描　　　述
-input /input/acled_cleaned/Nigeria_ACLED_cleaned.tsv \	MapReduce 输入数据的 HDFS 路径
-output /output/acled_analytic_out \	MapReduce 作业的 HDFS 输出路径
-mapper location_regains_mapper.py \	作为 map 函数的脚本运行,通过 STDIN/STDOUT 传递记录
-reducer location_regains_by_time.py \	作为 reduce 函数的脚本运行
-file location_regains_by_time.py \	当执行扩展的脚本,需要添加文件至分布式缓存
-file location_regains_mapper.py \	添加文件至分布式缓存
-jobconf stream.num.map.output.key.fields=2 \	对于 streaming 工具定义哪个或哪些字段作为 mapper 输出的键。例子中的 mapper 对每行记录输出三个字段。此参数设定程序将头两个字段作为键。这将影响 MapReduce 基于该复合字段对行记录的二次排序
-jobconf map.output.key.field.separator= \t \	设定键的分隔字符
-jobconf num.key.fields.for.partition=1 \	对于所有 map 的输出,如果键的第一个字段的值相等,则将其发送至相同的 reducer
-jobconf mapred.reduce.tasks=1	对输出键进行 reduce 操作的 JVM 任务数

在 map 阶段,Python 脚本获得与每行记录一致的行数据。调用 strip() 去除开头及结尾的空白字符,并按照制表符对行数据分隔。最终结果保存在行字段描述变量定义的数组中。

原始输入的 event_date 字段需要一些处理。为了允许框架对记录的时间进行升序排列,需要将当前的时间格式从 dd/mm/yy 转换为 yyyy-mm-dd。由于某些事件发生在 2000 年之前,需要将年份字段扩展为四位数字。一位数的天数以及月份都在开始位填 0 补齐,从而保证排序的正确性。

此分析工作只需要 location、event_date 以及 event_type 字段输出至 reduce 阶段。在 shell 脚本中，定义头两个字段作为输出键。定义 location 为第一个字段，将所有同一位置的记录分组至一个 reducer 中。定义 event_date 为第二个字段，允许 MapReduce 框架对 location、event_date 组成的复合字段排序。简单地将 event_type 字段作为每个键值对的值。

map 输出样式：

(cityA, 2010-08-09, explosion)
(cityB, 2008-10-10, fire)
(cityA, 2009-07-03, riots)

reducer 的排序部分将根据复合键 location、event_date 进行排序。

(cityA, 2009-07-03, riots)
(cityA, 2010-08-09,explosion)
(cityB, 2008-10-10,fire)

配置文件只定义了一个 reducer，因此本节中所有行划分至一个 Java 虚拟机（JVM）进行 reduce。如果定义为多 reduce 任务，cityA 以及 cityB 将会分配至不同的 JVM 进行 reduce 操作。

由于已经对数据按照位置进行分区划分，我们能保证每个划分的数据只会输入至相应的 mapper 中。此外，由于定义 event_date 作为排序列，能确保与给定地点相关的事件是按照时间升序排列的。至此，我们已经弄清楚整个脚本的工作原理了。

当 loc 的输入值发生变化时，脚本必须保持对其跟踪记录。由于位置数据已经排好序，这样的变化意味着已经完成对之前的位置的处理。将 current_loc 标志位为 START_OF_APP，并声明一个空的数组 govt_regains 存储我们感兴趣的事件时间。

程序首先将每行数据字段赋予各个变量。如果 loc 发生变化且非程序初始阶段，此时将当前的 govt_regains 集合输出至标准输出。该变化表明之前的位置已经处理完成，能正确地将时间序列集合从 reducer 输出。

如果输入的 loc 值与 current_loc 值相等，表明输入的事件仍然与当前处理的位置相关。判断事件类型是否为表示政府收复该领地的 regains 类型。如果是，将其加入至当前 govt_regains 集合。由于输入的记录根据 event_date 排序，可以保证添加至 govt_regains 的记录按照时间升序排列。

最终的结果是 reducer 输出的包含按照字典序排列的地点列表的单个文件。每个地点的右边部分为政府收复该地点的时间序列列表并以制表符分隔形式表示。

延伸阅读

Hadoop Streaming 是一个应用非常广泛的组件。下面是一些非常重要的补充知识。

使用任何支持标准输入/输出语言版本的 Hadoop Streaming

Hadoop Streaming 不仅仅支持 Python 版本,任何支持标准输入/输出的语言都能运行 Hadoop Streaming。我们常常通过 Java 类、shell 脚本、Ruby 脚本以及其他语言将已有的代码和功能改写为完整的 MapReduce 程序。

通过-file 参数为 MapReduce 作业传输需要添加的文件

与普通的 MapReduce 程序类似,可以使用分布式缓存传输应用额外需要的依赖文件。只需要简单的添加-file 参数。例如:

```
-file mapper.py \
-file wordlist.txt
```

3.10 在MapReduce中利用`MultipleOutputs`输出多个文件

用户使用 MapReduce 的一个常规需求是控制输出文件的名称而非简单地以 `part-*`命名。本节将展示如何使用类 `MultipleOutputs` 将不同的键值对输出至用户自定义的同一命名文件中。

准备工作

你需要从 `http://www.packtpub.com/support` 下载测试数据集 `ip-to-country.txt`,并将文件放置到 HDFS 中。

操作步骤

遵循如下步骤使用 `MultipleOutputs`。

1. 创建名为 `NamedCountryOutputJob` 的类,并配置 MapReduce 作业:

```
import org.apache.hadoop.conf.Configuration;
import org.apache.hadoop.fs.Path;
import org.apache.hadoop.io.IntWritable;
import org.apache.hadoop.io.LongWritable;
import org.apache.hadoop.io.NullWritable;
import org.apache.hadoop.io.Text;
import org.apache.hadoop.mapreduce.Job;
import org.apache.hadoop.mapreduce.Mapper;
import org.apache.hadoop.mapreduce.Reducer;
import org.apache.hadoop.mapreduce.lib.input.FileInputFormat;
import org.apache.hadoop.mapreduce.lib.input.TextInputFormat;
import org.apache.hadoop.mapreduce.lib.output.FileOutputFormat;
import org.apache.hadoop.mapreduce.lib.output.MultipleOutputs;
```

```java
import org.apache.hadoop.mapreduce.lib.output.TextOutputFormat;
import org.apache.hadoop.util.Tool;
import org.apache.hadoop.util.ToolRunner;

import java.io.IOException;
import java.util.regex.Pattern;

public class NamedCountryOutputJob implements Tool{
    private Configuration conf;
    public static final String NAME = "named_output";

    public static void main(String[] args) throws Exception {
        ToolRunner.run(new Configuration(), new NamedCountryOutputJob(), args);
    }
    public int run(String[] args) throws Exception {
        if(args.length != 2) {
            System.err.println("Usage: named_output <input> <output>");
            System.exit(1);
        }

        Job job = new Job(conf, "IP count by country to named files");
        job.setInputFormatClass(TextInputFormat.class);

        job.setMapperClass(IPCountryMapper.class);
        job.setReducerClass(IPCountryReducer.class);

        job.setMapOutputKeyClass(Text.class);
        job.setMapOutputValueClass(IntWritable.class);
        job.setJarByClass(NamedCountryOutputJob.class);

        FileInputFormat.addInputPath(job, new Path(args[0]));
        FileOutputFormat.setOutputPath(job, new Path(args[1]));

        return job.waitForCompletion(true) ? 1 : 0;
    }

    public void setConf(Configuration conf) {
        this.conf = conf;
    }

    public Configuration getConf() {
        return conf;
    }
}
```

2. 创建 mapper 输出键值对 counrty 以及数字 1：

```java
public static class IPCountryMapper
            extends Mapper<LongWritable, Text, Text, IntWritable>
{
        private static final int country_pos = 1;
        private static final Pattern pattern = Pattern.compile("\\t");

        @Override
```

```
            protected void map(LongWritable key, Text value,
                               Context context)   throws IOException, Interrupted
Exception {
                String country = pattern.split(value.toString())[country_pos];
                context.write(new Text(country), new IntWritable(1));
            }
        }
```

3. 创建用于统计每个 `country` 出现个数的 reducer，并使用 `MultipleOutputs` 分割输出文件：

```
public static class IPCountryReducer
            extends Reducer<Text, IntWritable, Text, IntWritable>
{
        private MultipleOutputs output;

        @Override
        protected void setup(Context context)
        throws IOException, Interrupted Exception {
            output = new MultipleOutputs(context);
        }

        @Override
        protected void reduce(Text key, Iterable<IntWritable> values,
Context context) throws IOException, InterruptedException
            {
            int total = 0;
            for(IntWritable value: values) {
                total += value.get();
            }
            output.write(new Text("Output by MultipleOutputs"),
                         NullWritable.get(), key.toString());
            output.write(key, new IntWritable(total), key.toString());
        }

        @Override
        protected void cleanup(Context context) throws IOException, Interrupted
Exception {
            output.close();
        }
    }
```

一旦作业成功运行，可以在给定的输出路径下面看到已命名的输出文件（如 `Qatar-r-#####`、`Turkey-r-#####`）。

工作原理

首先使用 Hadoop 提供的 `Tool` 接口创建该作业。类 `NamedCountryOutputJob` 中的 `run()` 方法用于检查提供的 HDFS 输入输出路径是否正确。此外，指定相应 mapper 类以及

reducer 类，并配置了用于读取文本行的 InputFormat 信息。

mapper 类定义了每一行数据中 `country` 字段出现位置的常量值。同时定义了用于分隔每行字段的正则表达。mapper 将每行出现的 `country` 作为键，数字 1 作为值输出。

在 reduce 阶段，每个 JVM 任务都将运行 `setup()` 方法，并初始化一个命名为 `output` 的 `MultipleOutputs` 实例。

每次调用 `reduce()` 输出一个 `country` 以及该 `country` 在数据集中出现的次数。使用一个计数器对次数进行求和。在最终的计数结果输出之前，需要使用输出实例写入一个头信息到对应的文件中。键的内容是头文本信息 `Output by MultipleOutputs`。由于不需要对应的输出值，因此值为空。根据 `key.toString()`，将相应头信息写入以当前 country 命名的用户自定义文件中。在下一行代码再次调用 `output.write()`，除了将输入键设置为输出键之外，还同时将最终的计数结果作为输出值，并且通过 `key.toString()` 方法指定输出文件输出，该文件与之前的 `output.write()` 方法的输出文件一致。

最终以 country 命名的文件既包含头信息也包含该 country 出现的统计次数。

使用 `MultipleOutputs`，不再需要在 job 设置过程中配置类 `OutputFormat`。并且，不限制 reducer 的输出键、值为具体的数据类型，可以分别输出数据类型为 Text/NullWritable 以及 Text/IntWritable 的键值对至同一文件。

3.11 创建用户自定义的 Hadoop Writable 及 InputFormat 读取地理事件数据

当从一个 MapReduce 应用读入或输出数据时，往往使用一个抽象类比 Hadoop 原生的 Writable 类（如 `Text`、`IntWritable`）更方便。本节介绍如何创建用户自定义的 Hadoop Writable 以及 InputFormat 供 MapReduce 使用。

准备工作

你需要从 `http://www.packtpub.com/support` 下载测试数据集 `Nigeria_ACLED_cleaned.tsv`，并将该文件放置到 HDFS 中。

操作步骤

遵循如下步骤，创建自定义的 InputFormat 以及 Writable 类。

1. 首先自定义两个 `WritableComparable` 类。这两个类描述如何在 mapper 端获取键值对，与 mapper 通过 `TextInputFormat` 获得 `LongWritable` 及 `Text` 方法类似。

简写键类：
```java
public class GeoKey implements WritableComparable {
    private Text location;
    private FloatWritable latitude;
    private FloatWritable longitude;
    public GeoKey() {
        location = null;
        latitude = null;
        longitude = null;
    }

    public GeoKey(Text location, FloatWritable latitude,
      FloatWritable longitude) {
        this.location = location;
        this.latitude = latitude;
        this.longitude = longitude;
    }

    //...getters and setters

    public void readFields(DataInput di) throws IOException {
        if (location == null) {
            location = new Text();
        }
        if (latitude == null) {
            latitude = new FloatWritable();
        }
        if (longitude == null) {
            longitude = new FloatWritable();
        }
        location.readFields(di);
        latitude.readFields(di);
        longitude.readFields(di);
    }
    public int compareTo(Object o) {
        GeoKey other = (GeoKey)o;
        int cmp = location.compareTo(other.location);
        if (cmp != 0) {
            return cmp;
        }
        cmp = latitude.compareTo(other.latitude);
        if (cmp != 0) {
            return cmp;
        }
        return longitude.compareTo(other.longitude);
    }

}
```

2. 编写对应的值类：
```java
public class GeoValue implements WritableComparable {
    private Text eventDate;
```

```java
    private Text eventType;
    private Text actor;
    private Text source;
    private IntWritable fatalities;

    public GeoValue() {
        eventDate = null;
        eventType = null;
        actor = null;
        source = null;
        fatalities = null;
    }

    //...getters and setters

    public void write(DataOutput d) throws IOException {
        eventDate.write(d);
        eventType.write(d);
        actor.write(d);
        source.write(d);
        fatalities.write(d);
    }

    public void readFields(DataInput di) throws IOException {
        if (eventDate == null) {
            eventDate = new Text();
        }
        if (eventType == null) {
            eventType = new Text();
        }
        if (actor == null) {
            actor = new Text();
        }
        if (source == null) {
            source = new Text();
        }
        if (fatalities == null) {
            fatalities = new IntWritable();
        }
        eventDate.readFields(di);
        eventType.readFields(di);
        actor.readFields(di);
        source.readFields(di);
        fatalities.readFields(di);
    }

    public int compareTo(Object o) {
        GeoValue other = (GeoValue)o;
        int cmp = eventDate.compareTo(other.eventDate);
        if (cmp != 0) {
            return cmp;
        }
        cmp = eventType.compareTo(other.eventType);
```

```
            if (cmp != 0) {
                return cmp;
            }
            cmp = actor.compareTo(other.actor);
            if (cmp != 0) {
                return cmp;
            }
            cmp = source.compareTo(other.source);
            if (cmp != 0) {
                return cmp;
            }
            return fatalities.compareTo(other.fatalities);
        }

    }
```

3. 接下来,创建用于序列化输入文件并生成 `GeoKey`、`GeoValue` 实例的 `InputFormat` 类。该 `InputFormat` 类继承于 Hadoop 提供的 `FileInputFormat` 类,返回一个自定义的 `RecordReader` 的实现。

```
public class GeoInputFormat extends FileInputFormat<GeoKey,
GeoValue> {

    @Override
    public RecordReader<GeoKey, GeoValue>
createRecordReader(InputSplit split, TaskAttemptContext context) {
        return new GeoRecordReader();
    }

    @Override
    protected boolean isSplitable(JobContext context, Path file) {
        CompressionCodec codec = new CompressionCodecFactory(context.
getConfiguration()).getCodec(file);
        return codec == null;
    }
}
```

4. 创建 `RecordReader` 用来读取 `Nigeria_ACLED_cleaned.tsv` 数据集。

```
public class GeoRecordReader extends RecordReader<GeoKey, GeoValue> {

    private GeoKey key;
    private GeoValue value;
    private LineRecordReader reader = new LineRecordReader();
    @Override
    public void initialize(InputSplit is, TaskAttemptContext tac)
throws IOException, InterruptedException {
        reader.initialize(is, tac);
    }

    @Override
    public boolean nextKeyValue() throws IOException,
InterruptedException {
```

```
            boolean gotNextKeyValue = reader.nextKeyValue();
            if(gotNextKeyValue) {
                if (key == null) {
                    key = new GeoKey();
                }
                if (value == null) {
                    value = new GeoValue();
                }
                Text line = reader.getCurrentValue();
                String[] tokens = line.toString().split("\t");
                key.setLocation(new Text(tokens[0]));
                key.setLatitude(new FloatWritable(Float.parseFloat(tokens[4])));
                key.setLongitude(new FloatWritable(Float.parseFloat(tokens[5])));

                value.setActor(new Text(tokens[3]));
                value.setEventDate(new Text(tokens[1]));
                value.setEventType(new Text(tokens[2]));
                try {
                    value.setFatalities(new IntWritable(Integer.parseInt(tokens[7])));
                } catch(NumberFormatException ex) {
                    value.setFatalities(new IntWritable(0));
                }
                value.setSource(new Text(tokens[6]));
            }
            else {
                key = null;
                value = null;
            }
            return gotNextKeyValue;
        }

        @Override
        public GeoKey getCurrentKey() throws IOException, InterruptedException {
            return key;
        }
        @Override
        public GeoValue getCurrentValue() throws IOException, InterruptedException {
            return value;
        }
        @Override
        public float getProgress() throws IOException, InterruptedException {
            return reader.getProgress();
        }
        @Override
        public void close() throws IOException {
            reader.close();
        }

}
```

5. 最后，为了测试 **InputFormat** 是否正确，创建一个只有 map 的作业。

```
public class GeoFilter extends Configured implements Tool {
```

```java
    public static class GeoFilterMapper extends Mapper<GeoKey, GeoValue,
Text, IntWritable> {
        @Override
        protected void map(GeoKey key, GeoValue value, Context context)
throws IOException, InterruptedException {
            String location = key.getLocation().toString();
            if (location.toLowerCase().equals("aba")) {
                context.write(value.getActor(),
                    value.getFatalities());
            }
        }
    }

    public int run(String[] args) throws Exception {

        Path inputPath = new Path(args[0]);
        Path outputPath = new Path(args[1]);

        Configuration conf = getConf();
        Job geoJob = new Job(conf);
        geoJob.setNumReduceTasks(0);
        geoJob.setJobName("GeoFilter");
        geoJob.setJarByClass(getClass());
        geoJob.setMapperClass(GeoFilterMapper.class);
        geoJob.setMapOutputKeyClass(Text.class);
        geoJob.setMapOutputValueClass(IntWritable.class);
        geoJob.setInputFormatClass(GeoInputFormat.class);
        geoJob.setOutputFormatClass(TextOutputFormat.class);

        FileInputFormat.setInputPaths(geoJob, inputPath);
        FileOutputFormat.setOutputPath(geoJob, outputPath);

        if(geoJob.waitForCompletion(true)) {
            return 0;
        }
        return 1;
    }
    public static void main(String[] args) throws Exception {
        int returnCode = ToolRunner.run(new GeoFilter(), args);
        System.exit(returnCode);
    }
}
```

工作原理

我们首先通过实现 `WritableComparable` 接口来自定义键和值。`WritableComparable` 接口提供了创建自定义抽象数据类型的能力,可以用来定义 MapReduce 框架中的键或值。

接下来，创建的 InputFormat 继承了类 `FileInputFormat`。Hadoop 提供的 `FileInputFormat` 是所有 InputFormat 的基础类。InputFormat 负责管理一个 MapReduce 作业的输入文件。因为我们不需要对输入文件的划分方式以及文件在集群中的传输方式进行改变，所以只需要重载 `createRecordReader()` 及 `isSplitable()` 这两个方法。

`isSplitable()` 方法是用来告诉类 `FileInputFormat`，输入的文件是否可以支持被分片，在 Hadoop 环境下是否存在可以读取并且分片输入文件的编码。`createRecordReader()` 用来创建 Hadoop 记录读取器，对每个独立的分片进行处理，并为 mapper 生成相应的键值对。

创建类 `GeoInputFormat` 之后，RecordReader 将处理每个输入文件分片，并为 mapper 创建 `GeoKey`、`GeoValue` 键值对作为输入。类 `GeoRecordReader` 重用 Hadoop 自带的类 `LineRecordReader` 的方法，从输入的 `split` 中读入数据。当 `LineRecordReader` 从数据集读入一条数据之后，我们创建两个对象，即 `GeoKey` 和 `GeoValue`，并将其传给 mapper。

第 4 章

使用 Hive、Pig 和 MapReduce 处理常见的任务

本章我们将介绍：
- 使用 Hive 将 HDFS 中的网络日志数据映射为外部表
- 使用 Hive 动态地为网络日志查询结果创建 Hive 表
- 利用 Hive 字符串 UDF 拼接网络日志数据的各个字段
- 使用 Hive 截取网络日志的 IP 字段并确定其对应国家
- 使用 MapReduce 对新闻档案数据生成 n-gram
- 通过 MapReduce 使用分布式缓存查找新闻档案数据中包含关键词的行
- 使用 Pig 加载一个表并执行包含 GROUP BY 的 SELECT 操作

4.1 介绍

在使用 Apache Hive、Pig 以及 MapReduce 的过程中，会发现有一些需要经常运行的任务。本章的内容对执行一些常见的工作流程提供了相应的解决方案。

这些工具会让你发现解决同样的问题可以有多种不同的方式，选择一个正确的实现是一个艰难的任务。下面的内容会使你的代码更加高效易读。

Hive 以及 Pig 为数据流与有意义的查询之间提供了清晰的抽象层，它们会将这些内容编译成复杂的 MapReduce 工作流。你可以利用 MapReduce 的强大功能对查询进

行扩展，而不必考虑底层的 MapReduce 语义。这些工具会解析相关表达并创建合适的 MapReduce 序列。Hive 使用了一种被称之为 **HiveQL** 的声明式类 SQL 语言，你可以通过这种语言对数据进行分析和关联。Pig Latin 书写的 Pig 操作是一种更紧凑精简的形式。

4.2 使用 Hive 将 HDFS 中的网络日志数据映射为外部表

我们时常需要为存在于 HDFS 的数据建立表，而这些数据又不是通过 Hive 数据仓库管理的。创建一个 Hive 外部表是处理这种情况最简单的方法。通常情况下，从 Hive 客户端查询外部表与查询 Hive 内部管理的表一样。

准备工作

请确认你可以通过已安装 Apache Hive 0.7.1 的客户机访问伪分布式 Hadoop 集群或完全分布式 Hadoop 集群，并配置好当前用户的环境路径。本节需要将 `weblog_entries` 数据集载入 HDFS 绝对路径 `/input/weblog/weblog_records.txt`。

操作步骤

执行下面的步骤，在 HDFS 中映射一个外部表。

1. 打开一个你常用的文本编辑器，最好具有 SQL 语法高亮功能。在本节，笔者本人使用一款名为 Textmate 的文本编辑器。

2. 添加如下 CREATE TABLE 语法：

```
DROP TABLE IF EXISTS weblog_entries;
CREATE EXTERNAL TABLE weblog_entries (
     md5 STRING,
     url STRING,
     request_date STRING,
     request_time STRING,
     ip STRING
)
ROW FORMAT DELIMITED FIELDS TERMINATED BY '\t' LINES TERMINATED BY '\n'
LOCATION '/input/weblog/';
```

3. 将脚本命名为 `weblog_create_external_table.hql` 并保存至工作路径。

4. 在 Hive 客户端通过操作系统 shell 命令，添加 `-f` 选项运行该脚本，命令如下：

```
hive -f weblog_create_external_table.hql
```

5. 将看见在 Hive 客户端显示两条成功的命令提示：

```
OK
Time taken: 2.654 seconds
OK
Time taken: 0.473 seconds
```

工作原理

首先，如果存在已定义的 `weblog_entries` 表结构，将其删除。接下来，使用带关键字 `EXTERNAL` 的 `CREATE` 语句告知 Hive 元数据存储：这些数据不被 HDFS 的 Hive 数据库管理。

表中的每个实体被定义为包含 5 个字段，即 URL 的 MD5 值、URL 本身、请求数据、请求的准确时间以及请求关联的 IP 地址。

行格式分隔符使用 Hive SerDe 的固有值。Hive SerDe 是 Hive 可扩展的并且内置的用于读取、记录原始数据的序列化反序列化机制。显式地声明 SerDe 使用制表符作为字段分隔符，换行符作为每行记录分隔符。当建立外部表时，Hive 需要使用 `LOCTATION` 关键字，用来指向包含表数据的 HDFS 绝对路径。

更多参考

当使用外部表时，下面有一些额外的技巧需要掌握。

LOCATION 必须指向一个目录而非文件

对于 0.7.1 版本，`LOCATION` 关键字需要指定 HDFS 的一个绝对路径。

删除外部表并没有删除存储在表中的数据

与 Hive 管理的表不同，`DROP` 命令只从元数据存储删除表对象实体而非对 HDFS 中的数据物理删除。依赖该 HDFS 路径下数据的其他应用在执行该操作后仍能照常工作。

将数据添加至 LOCATION 指定的路径下

如果新的数据被添加至外部表 `LOCATION` 属性指定的路径下，该数据将在后续对该表的查询结果中可见。

4.3　使用 Hive 动态地为网络日志查询结果创建 Hive 表

本节将介绍当查询执行的时候，创建内联表快速记录查询结果的技术。在进行查询之前定义每一个表是不切实际的，并且对于大规模 ETL 任务是缺乏可扩展性的。允许动态地定义临时表对于多步骤的复杂查询是非常有用的。

本节将创建一张新表，包含网络日志实体数据库的三个字段，分别是 `request_date`、`request_time` 以及 `url`。此外，定义了一个名为 `url length` 的新字段。

准备工作

请确认通过已安装 Apache Hive 0.7.1 的客户机能够访问伪分布式 Hadoop 集群或完全分布式 Hadoop 集群，并确认已经配置好当前用户的环境路径。

本节需要将 `weblog_entries` 数据集载入 Hive 表中，并且各个数据类型与下面的字段对应。

在 Hive 客户端输入命令：

```
describe weblog_entries
```

将看见如下的内容返回：

```
OK
md5 string
url string
request_date string
request_time string
ip string
```

操作步骤

执行下面的步骤，使用别名创建一个内联表。

1. 打开一个你常用的文本编辑器，最好具有 SQL 语法高亮功能。

2. 增加如下 CREATE TABLE 语法：

```
CREATE TABLE weblog_entries_with_url_length AS
SELECT url, request_date, request_time, length(url) as url_length
FROM weblog_entries;
```

3. 将脚本命名 `weblog_entries_create_table_as.hql` 为并保存至当前路径。

4. 在 Hive 客户端通过操作系统 shell 命令，添加 `-f` 选项运行该脚本，命令如下：

```
hive -f weblog_create_table_as.hql
```

5. 为了验证表已创建成功，直接在 Hive 客户端添加 `-e` 选项，执行如下命令：

```
hive -e "describe weblog_entries_with_url_length"
```

6. 你将看见一张表，包含三个 `string` 字段以及保存 URL 长度的第四个 `int` 字段：

```
OK
url string
```

```
    request_date string
    request_time string
    url_length int
```

工作原理

下面的语句初始化并命名定义一张新表 weblog_entries_with_url_length：

```
CREATE TABLE weblog_entries_with_url_length AS
```

之后将嵌套 SELECT 语句的结果集作为表别名的主体部分。在这种情况下，SELECT 语句从表 weblog_entries 的每行抓取字段 url、request_date 以及 request_time。字段的名称也复制至新表 weblog_entires_with_url_length 中。同时，定义一个新增的字段 url_length，类型为 int，用于计算、存储每行记录 url 字段的字符个数。

```
SELECT url, request_date, request_time, length(url) as url_length FROM
weblog_entries;
```

一个简单的语句生成的表既包含原始表的字段子集也包含新派生的字段。

更多参考

当使用外部表时，需要注意以下内容。

CREATE TABLE AS 不能用于生成外部表

对于 Apache Hive 0.7.1 版本，不能使用 SELECT 语句通过表别名创建外部表。

DROP 临时表

CREATE TABLE AS 语法非常简单，使得 Hive 的用户能非常容易地创建新表，但是不要忘记删除掉这些临时表。特别地，如果多次运行语句 CREATE ALIAS，后续执行的表将会因为表名冲突而失败。此外，每个临时表都会创建一个数据仓库命名空间，会不利于管理。

4.4 利用 Hive 字符串 UDF 拼接网络日志数据的各个字段

字符串拼接在任何开发任务中都是一种非常常见的操作。当使用 Hive 生成报表或简单的 ETL 任务时，该操作会经常出现。本节将展示使用一个 Hive 字符串连接 UDF 的基础且实用的例子。

本节我们将把表 weblog_entries 中分开的 request_date 和 request_time 字段进行合并。对于每行记录，在命令行输出单独的一列，其中包含字段 request_date 和

request_time，并以一个下划线（_）分割。

准备工作

请确认通过已安装 Apache Hive 0.7.1 的客户机能够访问伪分布式 Hadoop 集群或完全分布式 Hadoop 集群，并已配置好当前用户的环境路径。

本节需要将 weblog_entries 数据集载入 Hive 表中，并且将各个数据类型与下面的字段对应。

在 Hive 客户端输入命令：

```
describe weblog_entries
```

将看见如下的内容返回：

```
OK
md5 string
url string
request_date string
request_time string
ip string
```

操作步骤

执行下面的步骤，使用 HiveQL 对字符串进行连接。

1. 打开一个你常用的文本编辑器，最好具有 SQL 语法高亮功能。

2. 增加如下内联创建语法：

```
SELECT concat_ws('_', request_date, request_time) FROM weblog_entries;
```

3. 在当前目录保存并命名脚本为 weblog_concat_date_time.hql。

4. 在 Hive 客户端通过操作系统 shell 命令，添加 -f 选项运行该脚本。将在命令行显示 SELECT 的结果。下面的例子片段只包含两行实例。完整的输出一共包含 3000 行。

```
2012-05-10_21:33:26
2012-05-10_21:13:10
```

工作原理

该脚本的功能依赖于 Hive 内置的 UDF：将两个字符串进行连接并且指定连接符。对于每一行，分别以该行 request_date 和 request_time 的值作为函数的输入。函数的输出是一行字符串，包含上述两个字段以及字段分隔符，即下划线（_）。由于 SELECT 语句只包含该函数，且函数的输出为一行字符串，因此输出文件是包含一共 3000 行的字符串。

更多参考

下面的补充内容将有助于使用 `concat_ws()` 函数。

UDF `concat_ws()` 函数不会自动将参数转换为字符串

当传入 `concat_ws()` 的参数数据类型不是字符串时,将收到一条非常明确的错误信息:

```
FAILED: Error in semantic analysis: Line 1:21 Argument type mismatch
field1: Argument 2 of function CONCAT_WS must be "string", but "int"
was found.
```

如果希望自动将参数转换为字符串,可以使用常用的 `concat()` 函数。

为拼接后的字段取别名

与大多数 Hive UDF 一样,可以为 `concat_ws()` 的输出取别名。如果需要保存拼接后的结果并定义一个具有描述性的列名,这是十分有用的。

函数 `concat_ws()` 支持可变长度参数

使用 `concat_ws()` 至少需要在第一参数设定分隔字符,并输入一个需要输出的字符串参数。但是,并不限制输入字符串的参数个数。这些参数都会由分隔字符连接起来。

下面的用法是合法的:

`concat_ws('_','test')`

在命令行输出如下:

`test`

下面关于 `concat_ws()` 的使用同样是合法的:

`concat_ws('_','hi','there','my','name','is')`

在命令行输出如下:

`hi_there_my_name_is`

延伸阅读

第 6 章的如下章节与本节内容相关。

- 运用 Hive 日期 UDF 对地理事件数据集中的时间日期进行转换与排序(6.3 节)。
- 使用 Hive 创建基于地理事件数据的每月死亡报告(6.4 节)。

4.5 使用 Hive 截取网络日志的 IP 字段并确定其对应的国家

Hive 不直接支持外键。然而，根据同样的键值在一个表或多个表上对记录进行连接是非常常见的操作。本节将介绍如何根据表 `weblog_entries` 的请求 IP 字段将每一行请求日志与 IP 对应的国家进行简单的内连接。

对于表中的每一行记录，查询结果将在其尾部增加一个值，表明记录对应的国家。

准备工作

请确认通过已安装 Apache Hive 0.7.1 的客户机能够访问伪分布式 Hadoop 集群或完全分布式 Hadoop 集群，并配置好当前用户的环境路径。

本节需要将 `weblog_entries` 数据集载入 Hive 表中，并且各个数据类型与下面的字段对应。

在 Hive 客户端输入命令：

```
describe weblog_entries
```

将看见如下的内容返回：

```
OK
md5          string
url          string
request_date string
request_time string
ip           string
```

此外，本节需要将 `ip-to-country` 数据集载入对应的 Hive 表中，各个数据类型与下面的字段对应。

在 Hive 客户端输入命令：

```
describe ip_to_country
```

将看见如下的内容返回：

```
OK
Ip      string
country string
```

操作步骤

执行下面的步骤，使用 HiveQL 进行 Hive 表的内连接。

1. 打开一个你常用的文本编辑器，最好具有 SQL 语法高亮功能。

2. 添加如下内联创建语法：

```
SELECT wle.*, itc.country FROM weblog_entries wle
    JOIN ip_to_country itc ON wle.ip = itc.ip;
```

3. 当前目录保存并命名脚本为 `weblog_simple_ip_join.hql`。

4. 在 Hive 客户端通过操作系统 shell 命令，添加 -f 选项运行该脚本。将在命令行显示 SELECT 的结果。下面的例子片段只包含两行实例，完整的输出一共包含 3000 行。

```
11402ba8f780f7fbfb108f213449e1b9  /u.html  2012-05-10  21:19:05
98.90.200.33  United States
7ffb8f8ed136c5fe3a5dd6eedc32eae7  /cx.html  2012-05-10  21:17:05
59.19.27.24  Korea, Republic of
```

工作原理

语句 `SELECT wle.*` 会告诉 Hive 输出 `weblog_entires` 表中记录的每一列，其中 `wle` 是表名的一个简写别名。

此外，JOIN 操作等会告诉 Hive 每一行网络日志的记录 IP 地址，通过查找表 `ip_to_country` 确定其对应的国家。换言之，包含在这两张表的 IP 地址字段是连接键。

更多参考

下面提供的一些小技巧将更有助于使用 JOIN 语法。

Hive 支持多表连接

可以在一个 SELECT 语句内使用多个 JOIN <table> ON 实例与包含多个表的条件进行匹配。

用于内连接的 ON 操作不支持非等值连接

对于 Hive 0.7.1，不支持通过 ON 操作进行非等值条件的记录匹配。

一旦匹配条件变成非等值操作，本节同样的操作会失败。除了将每条记录与连接表中不匹配的 IP 记录做连接外，其他内容与之前的语句一样：

```
SELECT wle.*, itc.country FROM weblog_entries wle
    JOIN ip_to_country itc ON wle.ip != itc.ip;
```

该查询会产生错误：

```
FAILED: Error in semantic analysis: Line 2:30 Both left and right aliases encountered in JOIN ip
```

延伸阅读

本节是设计简单的表连接操作的一个参考。之后第 5 章的各节会覆盖更多复杂的 Hive Join。

- 使用 MapReduce 对数据进行连接（5.2 节）。
- 使用 Apache Pig 对数据进行复制连接（5.3 节）。
- 使用 Apache Pig 对有序数据进行归并连接（5.4 节）。
- 在 Apache Hive 中通过 map 端排序对地理事件进行分析（5.6 节）。

4.6 使用 MapReduce 对新闻档案数据生成 n-gram

n-gram 分析是查找全文文本词组块的一种方法，需要分析一组连续序列的词（gram）。本节将展示如何使用 Java MapReduce API 计算新闻文档数据的 n-gram。本节列出的一些代码对于各种不同的 MapReduce 作业也同样有用，包括：对于 `ToolRunner` 的设置，配置自定义参数以及自动移除之前作业提交的输出路径。

准备工作

本节假定你已经对 Hadoop 0.20 MapReduce API 有了大致的熟悉，且了解 n-gram 计算的基本原理。你需要读取本书提供的 `news_archives.zip` 数据集，该 ZIP 包含文件 `rural.txt` 以及 `science.txt`。将这两个文件放置于同一 HDFS 路径下面。

你需要使用 Hadoop 0.20 提供的 MapReduce API 在伪分布式集群或完全分布式集群上运行 MapReduce 作业。

同时需要将该代码打包成一个 JAR 文件，从而能在 shell 中通过 Hadoop JAR 启动器执行该程序。只需要编译 Hadoop 核心库并运行这个例子。

操作步骤

执行如下的步骤，在 MapReduce 中实现 n-gram。

1. 在 JAR 文件合适的源代码包中创建名为 `NGram.java` 的类。
2. 第一步涉及创建具体用于提交作业的 `Tool` 类，实现的方法如下：

```
import org.apache.hadoop.conf.Configuration;
import org.apache.hadoop.fs.FileSystem;
import org.apache.hadoop.fs.Path;
import org.apache.hadoop.io.LongWritable;
import org.apache.hadoop.io.NullWritable;
import org.apache.hadoop.io.Text;
import org.apache.hadoop.mapreduce.Job;
import org.apache.hadoop.mapreduce.Mapper;
import org.apache.hadoop.mapreduce.lib.input.FileInputFormat;
```

```
import org.apache.hadoop.mapreduce.lib.input.TextInputFormat;
import org.apache.hadoop.mapreduce.lib.output.FileOutputFormat;
import org.apache.hadoop.mapreduce.lib.output.TextOutputFormat;
import org.apache.hadoop.util.Tool;
import org.apache.hadoop.util.ToolRunner;

import java.io.IOException;
import java.util.regex.Pattern;

public class NGramJob implements Tool{

    private Configuration conf;

    public static final String NAME = "ngram";
    private static final String GRAM_LENGTH =  "number_of_grams";

    public void setConf(Configuration conf) {
        this.conf = conf;
    }

    public Configuration getConf() {
        return conf;
    }

    public static void main(String[] args) throws Exception {
        if(args.length != 3) {
            System.err.println("Usage: ngram <input> <output> <number_of_grams>");
            System.exit(1);
        }
        ToolRunner.run(new NGramJob(new Configuration()), args);
    }
    public NGramJob(Configuration conf) {
        this.conf = conf;
    }
```

3. run()方法用来设置输入输出格式、mapper 类的配置以及键值类的配置:

```
    public int run(String[] args) throws Exception {
        conf.setInt(GRAM_LENGTH, Integer.parseInt(args[2]));

        Job job = new Job(conf, "NGrams");
        job.setInputFormatClass(TextInputFormat.class);
        job.setOutputFormatClass(TextOutputFormat.class);
        job.setMapperClass(NGramJob.NGramMapper.class);
        job.setNumReduceTasks(0);
        job.setOutputKeyClass(Text.class);
        job.setOutputValueClass(NullWritable.class);
        job.setJarByClass(NGramJob.class);

        FileInputFormat.addInputPath(job, new Path(args[0]));
        FileOutputFormat.setOutputPath(job, removeAndSetOutput(args[1]));

        return job.waitForCompletion(true) ? 1 : 0;
    }
```

4. 方法 `removeAndSetOutput()` 不是必备的，但能避免输出目录已经存在的错误：

```
private Path removeAndSetOutput(String outputDir) throws IOException {
    FileSystem fs = FileSystem.get(conf);
    Path path = new Path(outputDir);
    fs.delete(path, true);
    return path;
}
```

5. `map()` 函数如下面的代码片段所示，通过继承 `mapreduce.Mapper` 实现：

```
public static class NGramMapper extends Mapper<LongWritable, Text, Text, NullWritable> {

    private int gram_length;
    private Pattern space_pattern = Pattern.compile("[ ]");
    private StringBuilder gramBuilder = new StringBuilder();

    @Override
    protected void setup(Context context) throws IOException, InterruptedException {
        gram_length = context.getConfiguration().getInt(NGramJob.GRAM_LENGTH, 0);
    }

    @Override
    protected void map(LongWritable key, Text value,Context context) throws IOException, InterruptedException {
        String[] tokens = space_pattern.split(value.toString());
        for (int i = 0; i < tokens.length; i++) {
            String token = tokens[i];
            gramBuilder.setLength(0);
            if(i + gram_length <= tokens.length) {
                for(int j = i; j < i + gram_length; j++) {
                    gramBuilder.append(tokens[j]);
                    gramBuilder.append(" ");
                }
                context.write(new Text(gramBuilder.toString()), NullWritable.get());
            }
        }
    }
}
```

工作原理

首先，设置需要载入的库，并创建名为 `NGram` 的公共类，实现 MapReduce 的 `Tool` 接口。静态变量 `NAME` 是有用的，需要在一个 Hadoop `Driver` 的实现设置这个值。`NGram` 程序需要按照特定的顺序传递三个参数：HDFS 的输入路径、HDFS 的输出位置以及计算每个词的 gram 总数。将类 `NGramJob` 的一个实例传给 `ToolRunner`，同时使用之前提到的参数

对 `Configuration` 对象进行初始化。

在 `run()` 方法中，配置作业使用 `TextInputFormat` 将文本的每行作为输入，使用 `TextOutputFormat` 将 map 过程中的每行文本作为输出。同时需要将 `Mapper` 类设置为公共的静态内部类 `NGramMapper`。由于这个作业是只有 map 的，我们将 reduce 的数设置为 0。之后设置 mapper 输出的键值对的 `Writable` 数据类型。调用 `setJarByClass()` 方法同样重要，这样 `TaskTrackers` 才能正确地解包并找到 `Mapper` 以及 `Reducer` 类。作业使用 `FileInputFormat` 以及 `FileOutputFormat` 的静态帮助方法设置输入输出路径。由于输出路径不允许存在，程序首先删除所有之前在给定路径下已存在的 HDFS 文件或目录。当所有的配置项正确之后，作业此时已准备好提交给 `JobTracker`。

类 `NGramMapper` 有一些非常重要的成员变量。变量 `gram_length` 引用自作业的配置项，是提交作业之前用户提供的参数。变量 `space_pattern` 是被静态编译的，是为了用空白字符进行正则分割。`gramBuilder` 是一个 `StringBulider` 实例，存储与每个字符串相关的由空格分隔的 gram 列表。mapper 接受的行号表示为一个 `LongWritable` 实例，该行内容表示为一个 `Text` 实例。函数立刻将该行分隔为空格分割的词组。对于每个词组，重新赋值给 `gramBuilder`，如果这个词组在该行的位置加上 `gram_length` 超过了该行的总长度，则忽略这个词组。否则，继续迭代并将后面的词组存入 `gramBuilder` 中直至循环次数达到 `gram_length`。之后输出 `gramBuilder` 的内容并进行外部的循环至下一个词组。最终的结果是存储在用户指定的输出路径的一个或多个部分文件，其中包含了由换行分隔列表构成的新闻档案的 *n*-gram。

输出的二元 gram（2 gram）的采样结果：

```
AWB has
has been
been banned
banned from
from trading
```

更多参考

下面的两部分讨论了如何有效地使用 `NullWritable` 对象以及提醒开发者谨慎地使用 HDFS 文件删除函数。

小心地调用 `FileSystem.delete()`

方法 `removeAndSetPath()` 在这个实现中自动地删除字符串参数给定的路径并且不产生任何警告。这个方法的参数是用户给定的输出路径，如果意外的将输入路径与输出路径颠倒了，则会删除输入路径。虽然这种在 MapReduce 的启动阶段动态地添加参数是十分方便的，但需要小心使用 FileSystem.delete()。

使用 `NullWritable` 避免不必要的序列化

程序使用 `NullWritable` 作为 mapper 的输出值的类型。由于程序对每一行只写单组 gram，只需要使用键输出我们的结果。如果 MapReduce 作业不需要同时输出键和值，使用 `NullWritable` 将避免框架对不需要的对象进行序列化输出至磁盘的问题。在大多数情况下，该方法往往是比使用空白占位符或输出静态单例更简洁，可读性更高。

4.7 通过 MapReduce 使用分布式缓存查找新闻档案数据中包含关键词的行

涉及调用第三方独立库和代码的复杂任务通常需要在 MapReduce 中使用分布式缓存。一个非常常见的操作是在每个 map/reduce 任务 JVM 中访问缓存的文件。本节将使用 MapReduce API 以及分布式缓存，标记在新闻档案数据集中包含在关键词列表中一个或多个词的数据行。使用分布式缓存确保每个 mapper 能获取到位于 HDFS 上的关键词列表。

准备工作

本节假定你已经对 Hadoop 0.20 MapReduce API 有了基本的了解。你需要访问本书提供的 `news_archives.zip` 数据集，该 ZIP 包含文件 `rural.txt` 以及 `science.txt`。分别将这两个文件放置同一 HDFS 路径下面。此外，ZIP 文件中包含了 `news_keywords.txt`。需要将该文件上传至 HDFS 绝对路径 `/cache_files/news_archives.txt`。同时支持在这个文件的新行处自由添加任意的新词。

你需要使用 Hadoop 0.20 提供的 MapReduce API 在伪分布式集群或完全分布式集群上运行 MapReduce 作业。

同时需要将该代码打包成一个 JAR 文件，从而能在 shell 中通过 Hadoop JAR launcher 执行该程序。只需要编译 Hadoop 核心库并运行这个例子。

操作步骤

下面的步骤实现了一个词匹配的 MapReduce 作业。

1. 在 JAR 文件合适的源代码包中创建名为 `LinesWithMatchingWordsJob.java` 的类。
2. 下面的代码提供一个用于提交作业的 `Tool` 类：

```
import org.apache.hadoop.conf.Configuration;
```

```java
import org.apache.hadoop.filecache.DistributedCache;
import org.apache.hadoop.fs.FileSystem;
import org.apache.hadoop.fs.Path;
import org.apache.hadoop.io.LongWritable;
import org.apache.hadoop.io.Text;
import org.apache.hadoop.mapreduce.Job;
import org.apache.hadoop.mapreduce.Mapper;
import org.apache.hadoop.mapreduce.lib.input.FileInputFormat;
import org.apache.hadoop.mapreduce.lib.input.TextInputFormat;
import org.apache.hadoop.mapreduce.lib.output.FileOutputFormat;
import org.apache.hadoop.mapreduce.lib.output.TextOutputFormat;
import org.apache.hadoop.util.Tool;
import org.apache.hadoop.util.ToolRunner;

import java.io.BufferedReader;
import java.io.File;
import java.io.FileReader;
import java.io.IOException;
import java.net.URI;
import java.util.HashSet;
import java.util.Set;
import java.util.regex.Pattern;

public class LinesWithMatchingWordsJob implements Tool {
    private Configuration conf;

    public static final String NAME = "linemarker";

    public void setConf(Configuration conf) {
        this.conf = conf;
    }

    public Configuration getConf() {
        return conf;
    }
    public static void main(String[] args) throws Exception {
        if(args.length != 2) {
            System.err.println("Usage: linemarker <input> <output>");
            System.exit(1);
        }
        ToolRunner.run(new LinesWithMatchingWordsJob(new Configuration()), args);
    }
    public LinesWithMatchingWordsJob(Configuration conf) {
        this.conf = conf;
    }
}
```

3. run()方法设置输入输出格式，完成 mapper 类的配置以及键值类的配置：

```java
public int run(String[] args) throws Exception {

    DistributedCache.addCacheFile(new Path("/cache_files/news_keywords.txt").toUri(), conf);
    Job job = new Job(conf, "Line Marker");
```

```
        job.setInputFormatClass(TextInputFormat.class);
        job.setOutputFormatClass(TextOutputFormat.class);
        job.setMapperClass(LineMarkerMapper.class);
        job.setNumReduceTasks(0);
        job.setOutputKeyClass(LongWritable.class);
        job.setOutputValueClass(Text.class);
        job.setJarByClass(LinesWithMatchingWordsJob.class);

        FileInputFormat.addInputPath(job, new Path(args[0]));
        FileOutputFormat.setOutputPath(job,new Path(args[1]));

        return job.waitForCompletion(true) ? 1 : 0;
    }
```

4. 下面的代码片段扩展 mapreduce.Mapper，实现 map() 方法:

```
    public static class LineMarkerMapper extends
Mapper<LongWritable, Text, LongWritable, Text> {

        private Pattern space_pattern = Pattern.compile("[ ]");
        private Set<String> keywords = new HashSet<String>();
```

5. 在 setup() 阶段，需要从分布式缓存读取并在本地磁盘写出该文件:

```
        @Override
        protected void setup(Context context) throws IOException,
InterruptedException {
            URI[] uris =DistributedCache.getCacheFiles(
             context.getConfiguration());
            FileSystem fs =
                FileSystem.get(context.getConfiguration());
            if(uris == null || uris.length == 0) {
               throw new IOException("Error reading file from
                       distributed cache. No URIs found.");
            }
            String localPath = "./keywords.txt";
            fs.copyToLocalFile(new Path(uris[0]), new
                             Path(localPath));
            BufferedReader reader = new BufferedReader(new
                             FileReader(localPath));
            String word = null;
            while((word = reader.readLine()) != null) {
                keywords.add(word);
            }
        }
```

map() 函数:

```
        @Override
        protected void map(LongWritable key, Text value,
                     Context context) throws
                        IOException, InterruptedException {
            String[] tokens =
                    space_pattern.split(value.toString());
            for(String token : tokens) {
```

```
                if(keywords.contains(token)) {
                    context.write(key, new Text(token));
                }
            }
        }
    }
}
```

工作原理

首先，设置需要载入的库，并创建名为 `LinesWithMatchingWordsJob` 的公共类，实现 MapReduce 的 `Tool` 接口。在作业提交之前，先检查输入、输出参数是否存在。在 `run()` 方法中，我们调用 `DistributedCache` 类的静态辅助方法 `addCacheFile()`，通过绝对路径 `/cache_files/news_keywords.txt` 的硬编码访问该 HDFS 缓存文件。文件包含的关键词由换行符分割，包含新闻语料库的兴趣词。将该路径对应的 `URI` 以及 `Configuration` 实例传给这个辅助方法。

接下来配置完成作业所需的其他配置项。由于需要处理文本，使用 `TextInputFormat` 和 `TextOutputFormat` 将每一行记录按照字符串读入和写出。同时需要将 `Mapper` 类设置为公共的静态内部类 `LineMarkerMapper`。由于该作业是只有 map 的，我们将 reduce 的数设置为 0。之后设置输出键的类型为 `LongWritable` 表示行数，输出值的类型为 `Text` 表示查找的关键词。调用 `setJarByClass()` 方法同样重要，以便 `TaskTrackers` 能正确地解包并找到 `Mapper` 以及 `Reducer` 类。该作业使用 `FileInputFormat` 和 `FileOutputFormat` 的静态辅助方法分别设置输入和输出路径。现在作业已配置完成，准备提交。

`Mapper` 类包含两个重要的成员变量。一个是用来以空格标记每行文本的静态编译的正则表达式，另一个是词表 `Set`，用于存储每个需要查找的兴趣词。

`Mapper` 中的 `setup()` 方法被告知获取位于当前分布式缓存的完整词表缓存文件的 URI。首先确认 URI 数组不为空，且包含的元素个数大于 0。如果数组满足上面的条件，获取位于 HDFS 的关键词文件，并将该文件写入一个临时工作目录中。保存内容为本地文件 `./keywords.txt`。此时，可以自由地使用标准 Java I/O 类在本地磁盘读写数据。文件中的每一行表示一个关键词，需要存储在一个关键词的 `HashSet` 中。在 `map()` 函数中首先通过空格符对一行记录分词，对于每个词查看是否包含于关键词列表中。如果匹配，输出发现该行的行号作为键，输出这个关键词作为值。

更多参考

下面提供的一些小技巧将有助于在 MapReduce 中使用分布式缓存。

使用分布式缓存将依赖的 JAR 发送至 map/reduce 任务 JVM。

map 和 reduce 任务时常依赖 JAR 文件形式的第三方库。如果将这些依赖放置在 HDFS，可以使用辅助的静态方法 `DistributedCache.addArchiveToClassPath()` 初始化包含依赖的作业。Hadoop 为作业的每个任务 JVM 添加该 JAR 文件作为依赖类路径。

分布式缓存在本地任务执行模式下不产生作用

如果将配置参数 `mapred.job.tracker` 设置为 `local`，`DistributedCache` 不能够用于配置在 HDFS 的归档文件或缓存文件。

4.8 使用 Pig 加载一个表并执行包含 GROUP BY 的 SELECT 操作

本节将使用 Pig 对包含在数据集 `ip_to_country` 中的 IP 地址进行分组并统计每个国家的 IP 地址数。

准备工作

请确认通过已安装了 Apache Pig 0.9.2 的客户机能够访问伪分布式 Hadoop 集群或完全分布式 Hadoop 集群，并确认当前用户账号的环境路径。本节需要将书中提供的 ip-to-country 数据集载入至 HDFS 绝对路径 `/input/weblog_ip/ip_to_country.txt`。

操作步骤

执行下面的步骤，在 Pig 中执行 SELECT 以及 GROUP BY 操作。

1. 打开一个你常用的文本编辑器，最好具有 SQL 语法高亮功能。

2. 添加如下的内联创建语法：

```
ip_countries = LOAD '/input/weblog_ip/ip_to_country.txt' AS
(ip: chararray, country:chararray);
country_grpd = GROUP ip_countries BY country;
country_counts = FOREACH country_grpd GENERATE FLATTEN(group),
COUNT(ip_countries) as counts;
STORE country_counts INTO '/output/geo_weblog_entries';
```

3. 将文件保存为 `group_by_country.pig`。

4. 在包含脚本的目录，执行使用添加 `-f` 选项的命令行调用 Pig 客户端。

工作原理

第一行根据存储在 HDFS 由制表符分隔的记录创建名为 `ip_countries` 的 Pig 关系。

这个关系定义了两个属性，分别为 ip 以及 country，其数据类型同为字符串数组。第二行创建了关系 country_grpd，每一行表示在关系 ip_countries 里各个不同的国家。第三行，Pig 会迭代关系 country_grpd，并计算在关系 ip_countries 中，与当前国家对应的记录行数。迭代的结果被保存为一个新的关系，名为 country_counts，其中包含的元组由两个属性构成，分别为 group 和 counts。存储包含在关系中的元组，并输出至指定路径 /output/geo_weblog_entries。

输出结果没有按照 country 升序或降序排序。

在 HDFS 路径/output/geo_weblog_entries 下，可以看到一个或多个包含由制表符分隔的国家列表，以及对应的 IP 地址数目。

延伸阅读

- ❑ 第 3 章的下列几节。
  - 使用 Apache Pig 过滤网络服务器日志中的爬虫访问量（3.3 节）。
  - 使用 Apache Pig 根据时间戳对网络服务器日志数据排序（3.4 节）。
- ❑ 第 6 章的使用 Pig 计算 Audioscrobbler 数据集中艺术家之间的余弦相似度（6.7 节）。

第 5 章

高级连接操作

本章我们将介绍:
- 使用 MapReduce 对数据进行连接
- 使用 Apache Pig 对数据进行复制连接
- 使用 Apache Pig 对有序数据进行归并连接
- 使用 Apache Pig 对倾斜数据进行倾斜连接
- 在 Apache Hive 通过 map 端排序对地理事件进行分析
- 在 Apache Hive 通过优化的全外连接分析地理事件数据
- 使用外部键值存储（Redis）连接数据

5.1 介绍

大多数处理环境需要将多个数据集进行连接后生成一些最后的结果。但是，在 MapReduce 进行连接操作是一个不寻常的并且代价非常高的任务。本章将介绍通过不同的方法在 Hadoop 进行数据连接，其中涉及的工具包括 MapReduce JAVA API、Apache Pig 和 Apache Hive。此外，本章将介绍如何利用外部存储资源使用 Hadoop MapReduce。

5.2 使用 MapReduce 对数据进行连接

MapReduce 中的数据连接是代价很高的操作。根据数据集的大小，可以选择在 map 端或者 reduce 端进行连接。在 map 端连接，两个或多个数据集在 MapReduce 作业的 map 阶段通过 key 进行连接。在 reduce 端连接，mapper 输出连接键，reduce 阶段负责连接这两个数据集。本节将介绍如何使用 MapReduce 在 map 端执行连接。我们将对一个网络日志数据集和一个 IP 与国家映射表进行关联。由于数据集将在 map 端进行关联，因此是一个只有 map 的作业。

准备工作

请从 http://www.packtpub.com/support 下载数据集 apache_nobots_tsv.txt 和 nobots_ip_country_tsv.txt，并载入 HDFS。

操作步骤

执行下面的步骤，使用 MapReduce 在 map 阶段连接数据。

1. 设置只有 map 的 MapReduce 作业，将 nobots_ip_country_tsv.txt 数据集载入至分布式缓存：

```
public class MapSideJoin extends Configured implements Tool {

  public int run(String[] args) throws Exception {

    Path inputPath = new Path(args[0]);
    Path outputPath = new Path(args[1]);

    Configuration conf = getConf();
    DistributedCache.addCacheFile(new
      URI("/user/hadoop/nobots_ip_country_tsv.txt"), conf);
    Job weblogJob = new Job(conf);
    weblogJob.setJobName("MapSideJoin");
    weblogJob.setNumReduceTasks(0);
    weblogJob.setJarByClass(getClass());
    weblogJob.setMapperClass(WeblogMapper.class);
    weblogJob.setMapOutputKeyClass(Text.class);
    weblogJob.setMapOutputValueClass(Text.class);
    weblogJob.setOutputKeyClass(Text.class);
    weblogJob.setOutputValueClass(Text.class);
    weblogJob.setInputFormatClass(TextInputFormat.class);
    weblogJob.setOutputFormatClass(TextOutputFormat.class);
    FileInputFormat.setInputPaths(weblogJob, inputPath);
    FileOutputFormat.setOutputPath(weblogJob, outputPath);

    if(weblogJob.waitForCompletion(true)) {
      return 0;
    }
    return 1;
  }
```

```java
public static void main( String[] args ) throws Exception {
    int returnCode = ToolRunner.run(new MapSideJoin(), args);
    System.exit(returnCode);
  }
}
```

2. 创建一个 mapper 从分布式缓存读取 nobots_ip_country_tsv.txt 数据集，并存储 IP/Country 表到一个 HashMap。

```java
public class WeblogMapper extends Mapper<Object, Text, Text, Text> {

  public static final String IP_COUNTRY_TABLE_FILENAME =
    "nobots_ip_country_tsv.txt";
  private Map<String, String> ipCountryMap = new
    HashMap<String, String>();

  private Text outputKey = new Text();
  private Text outputValue = new Text();

  @Override
  protected void setup(Context context) throws IOException,
InterruptedException {
      Path[] files = DistributedCache.getLocalCacheFiles(context.get
Configuration());
      for (Path p : files) {
        if (p.getName().equals(IP_COUNTRY_TABLE_FILENAME)) {
          BufferedReader reader = new BufferedReader(new FileReader
(p.toString()));
          String line = reader.readLine();
          while(line != null) {
            String[] tokens = line.split("\t");
            String ip = tokens[0];
            String country = tokens[1];
            ipCountryMap.put(ip, country);
            line = reader.readLine();
          }
        }
      }

    if (ipCountryMap.isEmpty()) {
      throw new IOException("Unable to load IP country table.");
    }
  }

  @Override
  protected void map(Object key, Text value, Context context) throws
IOException, InterruptedException {
      String row = value.toString();
      String[] tokens = row.split("\t");
      String ip = tokens[0];
      String country = ipCountryMap.get(ip);
```

```
            outputKey.set(country);
            outputValue.set(row);
            context.write(outputKey, outputValue);
        }
    }
```

3. 运行该作业：

```
$ hadoop jar AdvJoinChapter5-1.0.jar com.packt.ch5.advjoin.mr.MapSideJoin
/user/hadoop/apache_nobots_tsv.txt /user/hadoop/data_jnd
```

工作原理

在第一步中，调用静态方法：

```
DistributedCache.addCacheFile(new URI("/user/hadoop/nobots_ip_country_tsv.txt"),
conf)
```

该方法将设置作业配置项中的 `mapred.cache.files` 属性。`mapred.cache.files` 属性用于告知 MapReduce 框架将文件 `nobots_ip_country_tsv.txt` 分配至每一个要启动 mapper 的集群节点（也可以包括 reducer，如果作业配置为需要运行 reducer 的）。

在第二步中，我们重载了 mapper 的 `setup()` 方法。MapReduce 框架对于每次任务只调用一次 `setup()` 方法，其优先级高于其他的 `map()` 方法。`setup()` 方法是对 mapper 类进行一次性初始化操作最好的地方。

为了读取分布式缓存，需要调用静态方法 `DistributedCache.getLocalCacheFiles(context.getConfiguration())` 获得放置在分布式缓存的所有文件。接下来，遍历所有放置在分布式缓存的文件，其中只有一个是 `nobots_ip_country_tsv.txt` 数据集。将其数据加载至一个 HashSet 中。

最后，在 `map()` 方法中，使用在 `setup()` 方法加载的 HashSet，根据文件 `apache_nobots_tsv.txt` 中与每个 IP 关联的国家值连接 `nobots_ip_country_tsv.txt` 与文件 `apache_nobots_tsv.txt`。

更多参考

MapReduce 框架同样支持使用分布式缓存分发归档文件。归档文件可以是一个 ZIP 文件、GZIP 文件，甚至可以是 JAR 文件。一旦归档文件被分发至各个任务节点，这些文件将会被自动解压。

为了向分布式缓存增加一个归档文件，可以简单地在配置 MapReduce 作业的时候使用类 `DistributedCache` 的静态方法 `addCacheArchive()`：

```
DistributedCache.addCacheArchive(new URI("/user/hadoop/nobots_ip_country_
tsv.zip"), conf);
```

延伸阅读

- 使用 Apache Pig 对数据进行复制连接（5.3 节）。
- 使用 Apache Pig 对有序数据进行归并连接（5.4 节）。
- 使用 Apache Pig 对倾斜数据进行倾斜连接（5.5 节）。

5.3 使用 Apache Pig 对数据进行复制连接

Apache Pig 提供了多种高级连接方法，包括：

- Reduce 端连接；
- 复制连接；
- 合并连接；
- 倾斜连接。

当使用 Pig 的 JOIN 操作时，reduce 端连接是默认的实现方式。如果指定 replicated 或 merge 关键词，Pig 也同样支持 map 端的连接。本节将介绍如何使用 Pig 在 map 端进行复制连接。我们将对一个网络日志数据集和一个 IP 与国家映射表进行关联。

准备工作

从 http://www.packtpub.com/support 下载 apache_nobots_tsv.txt 以及 nobots_ip_country_tsv.txt 数据集，并放至 HDFS 上。同时，需要在集群上安装最新版本（0.9 或以上）的 Apache Pig。

操作步骤

执行下面的步骤，在 Apache Pig 中完成复制连接。

1. 打开你常用的文本编辑器，创建文件 replicated_join.pig。生成两个 Pig 关系[①] 用于读取两个数据集：

```
nobots_weblogs = LOAD '/user/hadoop/apache_nobots_tsv.txt' AS
(ip: chararray, timestamp:long, page:chararray, http_status:int, payload_
```

[①] Pig 中的一个关系相当于关系数据库中的一个表：http://pig.apache.org/docs/ro.12.0/basic.htm/#relations。——译者注

```
size:int, useragent:chararray);
    ip_country_tbl = LOAD '/user/hadoop/nobots_ip_country_tsv.txt' AS (ip:chararray,
country:chararray);
```

2. 使用关键词 `replicated` 连接两个数据集：

```
weblog_country_jnd = JOIN nobots_weblogs BY ip, ip_country_tbl BY ip USING
'replicated';
```

3. 格式化连接后的结果并存储该结果：

```
cleaned = FOREACH weblog_country_jnd GENERATE ip_country_tbl::ip, country,
timestamp, page, http_status, payload_size, useragent;
    STORE cleaned  INTO '/user/hadoop/weblog_country_jnd_replicated';
```

4. 执行该作业：

```
$ pig -f replicated_join.pig
```

工作原理

第一步，定义了两个关系 `nobots_weblogs` 以及 `ip_country_tbl`，分别对应两个输入数据集。第二步，使用 Pig 的复制连接根据 `ip` 字段对两个数据集进行连接。Pig 将把最右侧的关系，即 `ip_country_tbl`，放入内存，并与关系 `nobots_weblogs` 进行连接。这里很重要的是，最右侧的关系大小要足够小以保证能载入 mapper 端的内存。如果数据集太大，Pig 不会产生警告，而是会产生内存不足的异常，导致作业失败。

第三步，格式化连接后的关系至一个新的名为 `cleaned` 的关系。在 FOREACH 语句中有一个字段看起来比较异样，那就是 `ip_country_tbl::ip`。由于连接之后的关系包含两个同名的字段 `ip`，需要使用::运算符定义在关系中希望存储哪一列数据。我们也可以使用 `nobots_weblogs::ip` 轻易地进行替换，在这个例子里没有任何区别。

更多参考

复制连接能用于多个关系的连接。例如，修改上面的内容，使用复制连接对 3 个关系进行内部连接：

```
weblog_country_jnd = JOIN nobots_weblogs BY ip, ip_country_tbl BY ip,
another_relation BY ip USING 'replicated';
```

同样的，右侧的数据集必须装载至内存。在这个例子中，`ip_country_tbl` 以及 `another_relation` 都需要载入 mapper 的内存。

延伸阅读

- 使用 Apache Pig 对有序数据进行归并连接（5.4 节）。
- 使用 Apache Pig 对倾斜数据进行倾斜连接（5.5 节）。

5.4 使用 Apache Pig 对有序数据进行归并连接

与上一节提到的复制连接类似，Apache Pig 的归并连接是另一种 map 端的连接技术。然而，这两种实现主要的不同是：归并连接不需要将任何数据放入内存。本节将展示如何使用 Pig 对两个数据集进行归并连接。

准备工作

从 http://www.packtpub.com/support 下载 apache_nobots_tsv.txt 以及 nobots_ip_country_tsv.txt 数据集，并载入至 HDFS 上。同时，需要在集群上安装最新版本（0.9 或以上）的 Apache Pig。

为了在 Pig 中使用归并连接，两个数据集需要根据连接键进行排序。为了排序两个数据集，可以执行如下的 Unix sort 命令：

```
$ sort -k1 apache_nobots_tsv.txt > sorted_apache_nobots_tsv.txt
$ sort -k1 nobots_ip_country_tsv.txt > sorted_nobots_ip_country_tsv.txt
```

将排序好的文件放在 HDFS 上：

```
$ hadoop fs -put sorted_apache_nobots_tsv.txt /user/hadoop
$ hadoop fs -put sorted_nobots_ip_country_tsv.txt /user/Hadoop
```

操作步骤

执行下面的步骤，在 Apache Pig 中完成归并连接。

1. 打开文本编辑器，创建文件 merge_join.pig。生成两个 Pig 关系用于读取两个数据集：

```
nobots_weblogs = LOAD '/user/hadoop/sorted_apache_nobots_tsv.txt' AS (ip: chararray, timestamp:long, page:chararray, http_status:int, payload_size:int, useragent:chararray);
ip_country_tbl = LOAD '/user/hadoop/sorted_nobots_ip_country_tsv.txt' AS (ip:chararray, country:chararray);
```

2. 使用关键字 merge 连接两个数据集：

```
weblog_country_jnd = JOIN nobots_weblogs BY ip, ip_country_tbl BY ip USING 'merge';
```

3. 格式化连接后的结果并存储该结果：

```
cleaned = FOREACH weblog_country_jnd GENERATE ip_country_tbl::ip, country, timestamp, page, http_status, payload_size, useragent;
STORE cleaned INTO '/user/jowens/weblog_country_jnd_merge';
```

4. 执行该作业：
```
$ pig -f merge_join.pig
```

工作原理

第一步，定义了两个关系 `nobots_weblogs` 以及 `ip_country_tbl`，分别对应两个输入数据集。

第二步，使用 Pig 的归并连接根据 `ip` 字段对两个数据集进行连接。Pig 将会启动两个 MapReduce 作业执行归并连接。首先，Pig 将发送所有与 `nobots_weblogs` 关系关联的数据至所有的 mapper，并采样 `ip_country_tbl` 数据建立一个索引。关于 JOIN 语句，将两个关系中数据量大的一个放在左侧输入是很重要的，正如将 `nobots_weblogs` 放在左侧一样。一旦 Pig 创建了索引，将启动第二个只有 map 的作业，读入左侧的关系以及在第一个 MapReduce 作业建立的索引，连接这两个关系。

更多参考

Pig 的归并连接需要输入文件的数据按照升序排序是至关重要的。此外，所有输入数据必须根据文件名升序排序。例如，如果关系 `nobots_weblogs` 在两个输入文件包括 3 个不同的 IP，这些 IP 会以如下的形式分布。

- 名为 part-00000 的文件包含 IP 为 111.0.0.0 的行。
- 在名为 part-00000 的文件中，包含 IP 为 112.0.0.0 的行必须出现在行 111.0.0.0 的后面。
- 包含 IP 为 222.0.0.0 的行出现在名为 part-00001 的文件中。

本示例表明可以对多个文件中的 IP 按照名字进行整体排序。文件名需要升序，因为 Pig 默认按照升序的文件名读取已排好序的数据。

延伸阅读

- 使用 Apache Pig 对倾斜数据进行倾斜连接（5.5 节）。

5.5 使用 Apache Pig 对倾斜数据进行倾斜连接

在分布式处理环境中，如果在 map 阶段不能均匀地分割输出键元组，将发生严重的数据倾斜问题，这将导致不一致的处理时间。在 MapReduce 框架中，数据倾斜将导致在同一作业中一些 mapper/reducer 将明显花费比其他 mapper/reducer 更多的时间处理一个任务。

Apache Pig 利用倾斜连接解决数据倾斜带来的连接问题。本节将展示如何将一个倾斜数据集与一个小表进行连接。

准备工作

从 http://www.packtpub.com/support 下载 apache_nobots_tsv.txt 以及 nobots_ip_country_tsv.txt 数据集，并载入至 HDFS 上。同时，需要在集群上安装最新版本（0.9或以上）的 Apache Pig。

为了倾斜文件 skewed_apache_nobots_tsv.txt 中的数据，创建如下 shell 脚本，添加同一行记录上千次至新文件 skewed_apache_nobots_tsv.txt 中：

```
#!/bin/bash

cat apache_nobots_tsv.txt > skewed_apache_nobots_tsv.txt
for i in {1..5000}
do
  head -n1 apache_nobots_tsv.txt >> skewed_apache_nobots_tsv.txt
done
```

在文件 skewed_apache_nobots_tsv.txt 中，IP 地址 221.220.8.0 出现的次数明显比其他的 IP 高。

将文件 skewed_apache_nobots_tsv.txt 以及 nobots_ip_country_tsv.txt 加载至 HDFS 上：

```
$hadoop fs -put skewed_apache_nobots_tsv.txt /user/hadoop/
$hadoop fs -put nobots_ip_country_tsv.txt /user/hadoop/
```

操作步骤

执行下面的步骤，在 Apache Pig 中完成归并连接。

1. 打开文本编辑器，创建文件 skewed_join.pig。生成两个 Pig 关系用于读取两个数据集：

```
nobots_weblogs = LOAD '/user/hadoop/sorted_apache_nobots_tsv.txt' AS (ip: chararray, timestamp:long, page:chararray, http_status:int, payload_size:int, useragent:chararray);
ip_country_tbl = LOAD '/user/hadoop/nobots_ip_country_tsv.txt' AS (ip:chararray, country:chararray);
```

2. 使用关键词 skewed 连接两个数据集：

```
weblog_country_jnd = JOIN nobots_weblogs BY ip, ip_country_tbl BY ip USING 'skewed';
```

3. 格式化连接后的结果并存储该结果：

```
cleaned = FOREACH weblog_country_jnd GENERATE ip_country_tbl::ip,country,
```

```
timestamp, page, http_status, payload_size, useragent;
    STORE cleaned INTO '/user/hadoop/weblog_country_jnd_skewed';
```

4. 执行该作业：

```
$ pig -f skewed_join.pig
```

工作原理

第一步，定义了两个关系 `nobots_weblogs` 以及 `ip_country_tbl`，分别对应两个输入数据集。

第二步，使用 Pig 的倾斜连接根据 `ip` 字段对两个数据集进行连接。Pig 将会启动两个 MapReduce 作业执行倾斜连接。第一个 MapReduce 作业将会对 `nobots_weblogs.txt`（倾斜数据）进行采样。第二个 MapReduce 作业将执行一个 reduce 端的连接操作。Pig 将根据第一个 MapReduce 作业的采样结果决定如何将数据分配至 reducer。如果在当前数据集出现数据倾斜，Pig 将试图优化到 reducer 的数据分配。

5.6 在 Apache Hive 中通过 map 端连接对地理事件进行分析

在 Apache Hive 中连接两个表，其中的一个表可能明显小于另一个表。在这种情况下，Hive 会以散列表（hash table）的形式表示小表，并将其推送至分布式缓存，在 map 端完成整个连接表的操作。在本节，我们使用 map 端连接对任何重要的节日信息以及当天可能发生的地理事件进行关联。

准备工作

确保 Apache Hive 0.7.1 已安装至你的客户端，并保证当前用户的环境路径正确。

本节需要将数据集 `Nigera_ACLED_cleaned.tsv` 加载为 Hive 表 `acled_nigeria_cleaned`，且下面的字段分别对应正确的数据类型。可以从 http://www.packtpub.com/support 下载数据集 `Nigera_ACLED_cleaned.tsv`。

在 Hive 客户端输入命令：

```
describe acled_nigeria_cleaned
```

将返回如下结果：
```
OK
loc         string
event_date  string
event_type  string
```

```
actor       string
latitude    double
longitude   double
source      string
fatalities  int
```

本节同样需要将数据集 `nigeria-holidays.tsv` 加载为 Hive 表 `nigeria_holidays`，且下面的字段分别对应正确的数据类型。

在 Hive 客户端输入命令：

```
describe nigeria_holidays
```

将返回如下结果：

```
OK
yearly_date   string
description   string
```

操作步骤

执行下面的步骤，在 Apache Hive 中完成 map 端连接。

1. 打开文本编辑器，创建文件 `map-join-acled-holiday s.sql`。

2. 添加内联创建、转换语义。

```sql
SELECT /*+ MAPJOIN(nh)*/ acled.event_date, acled.event_type,
nh.description
    FROM acled_nigeria_cleaned acled
    JOIN nigeria_holidays nh
        ON (substr(acled.event_date, 6) = nh.yearly_date);
```

3. 在系统 shell 添加 Hive 客户端 `-f` 选项执行脚本。如果在输出 trace 看见如下消息，表明 map 端连接正在执行：

```
Mapred Local Task Succeeded. Convert the Join into MapJoin
```

生成的 MapReduce 作业不会包含任何 reduce 任务。

4. 在输出命令行将首先出现如下几行内容：

```
2002-01-01  Riots/Protests  New Years Day
2001-06-12  Battle-No change of territory  Lagos State only; in
memory of failed 1993 election
2002-05-29  Violence against civilians  Democracy Day
2010-10-01  Riots/Protests  Independence Day
2010-10-01  Violence against civilians  Independence Day
```

工作原理

脚本对表 `acled_nigeria_cleaned` 的列 `event_date` 中的月-日部分与表 `nigeria-`

holidays 的列 `yearly_date` 做连接。`substr(event_date, 6)` 将从第六个字符开始，忽略在列的每行记录的年部分。内部提示 `/*+ MAPJOIN(nh) */`，人工的定义了载入每个 mapper 的小表。表 `nigeria_holidays` 是非常小的，将其作为散列表是非常合理的。每个 map 的 join 过程将表 `acled_nigeria_cleaned` 中的每一行与自己复制的散列表 `nigeria_holidays` 连接。MAPJOIN 操作生成了散列表并且将其分配至每个 map 任务。

最后，将 event_date、event_type 列的值以及事件发生假日的描述作为结果。

更多参考

map 端的连接有一些小技巧以及有用的配置项。这里讲述了一些内容。

在可能的情况，自动转换为 map 端连接

在 Hive 配置项中将 `hive.auto.convert.join` 设置为 true，只要表的大小低于一个阈值，Hive 会试图将连接转化为 map 端连接。可以配置最大阈值属性 `hive.smalltable.filesize`，该属性决定 Hive 将多大（或更小）的文件作为小表。这个值用 long 类型表示字节数（比如，25000000L = 25M）。

同样可以考虑设置 `hive.hashtable.max.memory.usage`。如果 map 任务需要的内存超过配置的百分比，map 任务将会终止。

Map 端 join 的行为

如果忽略了 `/*+ MAPJOIN() */` 而依赖自动转化，跟踪 Hive 的 join 优化行为是一件比较困难的事。以下是一些小技巧。

- `TableFoo LEFT OUTER JOIN TableBar`：试图转化 `TableBar` 为一个散列表。
- `TableFoo RIGHT OUTER JOIN TABLE B`：试图转化 `TableFoo` 为一个散列表。
- `TableFoo FULL OUTER JOIN TableBar`：框架不会将全外连接转化为 map 端连接。

延伸阅读

- 在 Apache Hive 通过优化的全外连接分析地理事件数据（5.7 节）。

5.7 在 Apache Hive 通过优化的全外连接分析地理事件数据

本节将尼日利亚 VIP 的列表与任何 VIP 的生日发生的尼日利亚 ACLED 事件进行关联。我们不但对事件是否发生在名人生日当天感兴趣，也对那些与任何事件都不发生关系

的人感兴趣。在单个查询中完成这样的分析，全外连接是最有意义的。同时需要将结果存入一个表中。

准备工作

确保 Apache Hive 0.7.1 已安装至你的客户端，并保证当前用户的环境路径正确。

本节需要将数据集 Nigera_ACLED_cleaned.tsv 加载为 Hive 表 acled_nigeria_cleaned，且下面的字段分别对应正确的数据类型。可以从 http://www.packtpub.com/support 下载数据集 Nigera_ACLED_cleaned.tsv。

在 Hive 客户端输入下面的命令：

```
describe acled_nigeria_cleaned
```

将返回如下结果：

```
OK
loc        string
event_date string
event_type string
actor      string
latitude   double
longitude  double
source     string
fatalities int
```

本节同样需要将数据集 nigeria-vip-birthdays.tsv 加载为 Hive 表 nigeria_vips，且下面的字段分别对应正确的数据类型。可以从 http://www.packtpub.com/support 下载数据集 nigeria-vip-birthdays.tsv。

在 Hive 客户端输入下面的命令：

```
describe nigeria_ vips
```

将返回如下结果：

```
OK
name        string
birthday    string
description string
```

操作步骤

执行下面的步骤，在 Apache Hive 中完成全外连接。

1. 打开文本编辑器，创建文件 full_outer_join_acled_vips.sql。
2. 添加内联创建、转换语义：

```
DROP TABLE IF EXISTS acled_nigeria_event_people_links;
CREATE TABLE acled_nigeria_event_people_links AS
SELECT acled.event_date, acled.event_type, vips.name, vips.
description as pers_desc, vips.birthday
    FROM nigeria_vips vips
    FULL OUTER JOIN acled_nigeria_cleaned acled
        ON (substr(acled.event_date,6) = substr(vips.birthday,
6));
```

3. 在系统 shell 添加 Hive 客户端 `-f` 选项执行脚本 `full_outer_join_acled_vips.sql`。

4. 一旦脚本执行完成，将显示一共 2931 条记录载入至表 `acled_nigeria_event_people_links` 中。

5. 在 Hive 的 shell 中执行如下查询：

```
SELECT * FROM acled_nigeria_event_people_links WHERE event_date IS
NOT NULL AND birthday IS NOT NULL limit 2";
```

6. 将看见如下输出：

```
OK
2008-01-01  Battle-No change of territory  Jaja  Wachuku "First
speaker of the Nigerian House of Representatives"  1918-01-01

2002-01-01  Riots/Protests  Jaja Wachuku  "First speaker of the
Nigerian House of Representatives"  1918-01-01
```

工作原理

首先，删除之前已存在被命名为 `acled_nigeria_event_people_links` 的表。使用内联语句 `CREATE TABLE AS`，快捷地显式定义了表。

全外连接会匹配来自 `acled_nigeria_cleaned` 的行以及来自 `nigeria_vips` 的行，其中列 `event_date` 以第六个位置开始的子字符串与 VIP 的生日相等。使用方法 `substr(event_date, 6)` 消除记录中列 `event_date` 的年份作为比较的部分。

获得的表中的行包括从 `SELECT` 语句取得的 `acled.event_date`、`acled.event_type`、`vips.name`、`vips.description as pers_desc` 以及 `vips.birthday`。列 `vips.description` 取了个别名 `pers_desc`，使列标签更具意义。如果事件记录没有与任何生日匹配，列 `vips.name`、`vips.description` 以及 `pers_desc` 将为 NULL。对于没有匹配事件的人，列 `acled.event_date` 和列 `acled.event_type` 也将为空。

为了优化 reducer 的吞吐量，将表 `nigeria_vips` 写在 `FROM` 处，表 `acled_nigeria_cleaned` 写在 `JOIN ON` 子句中。由于运行的是常规的 Hive 连接而非 map 端的连接，实际

的表连接操作将发生至 reduce 端。Hive 将会尝试缓存左侧的表的行，传输右侧的表。表 `nigeria_vips` 比表 `acled_nigeria_cleaned` 小很多，通过设计查询语法，使得传输表 `acled_nigeria_cleaned` 的行记录且缓存 `nigeria_vips` 中的记录，从而达到减少使用 reducer 内存空间的目的。

实际上，对于这个特殊的 VIP 列表而言，不存在生日当天不发生位于 `acled_nigeria_cleaned` 列表的事件。因此，外连接未产生一行生日未匹配任意事件的情况。此外，在列表中没有两个人的生日相同。因此，外连接不会将同一事件与多个 VIP 的生日关联。结果表中包含 2931 行记录，与表 `acled_nigeria_cleaned` 的行数完全一致。

更多参考

下面是其他的一些改进 Hive 中连接操作性能的东西。

常规连接与 map 端连接

Hive 文档中提到的"常规连接"指在 reducer 端需要对行数据进行物理连接。map 端连接，顾名思义，通过并行的 map 任务进行连接操作，省去了不必要的 reduce 过程。

STREAMTABLE 提示

使用 `/*+ STREAMTABLE(tablename) */`，可以指定在 reduce 过程中哪个表作为流传输。

在查询语句中表的顺序

在查询中需要连接的表的左右顺序，特别是多表连接的情况下，是非常重要的。Hive 尝试将左侧的表的行记录进行缓存，右侧的表的结果作为流传输。在多表连接中，可能会有多个 map/reduce 任务，然而执行的原理是相同的。第一个连接的结果会被缓存起来，而接下来的右侧表的行数据是作为流的。使用上面的原理能明智地调整表的连接顺序。

5.8　使用外部键值存储（Redis）连接数据

键值存储是存储大数据有效的工具。在 MapReduce 中，可以使用键值存储来存储那些不能存储在一个或多个 mapper（提示：多个 mapper 会运行在同一个 slave 节点）内存却能存储在服务器内存的大数据集。

本节将介绍如何使用 Redis 执行 MapReduce 的 map 端连接操作。

准备工作

首先，下载并安装 Redis。本书使用 Redis 2.4.15 版本。在 Redis 的网站，可以得到快速指南手册 http://redis.io/topics/quickstart。一旦编译并安装好 Redis 服务器，使用如下的命令行将启动服务器：

```
$ redis-server
```

验证 Redis 服务器正常工作可以使用 redis-cli：

```
$ redis-cli ping
```

如果服务已经正常启动，Redis 会返回消息 "PONG"。

接下来，需要从 https://github.com/xetorthio/jedis 下载并编译 Jedis。Jedis 是 Redis 的 Java 客户端，在 MapReduce 程序中使用该客户端与 Redis 通信。本书使用的 Jedis 版本为 2.1.0。

最后，从 http://www.packtpub.com/support 下载数据集 apache_nobots_tsv.txt 以及 nobots_ip_country_tsv.txt。将 apache_nobots_tsv.txt 放到 HDFS 上，并将文件 nobots_ip_country_tsv.txt 留在当前的工作文件夹中。

操作步骤

按照下面的步骤，使用 Redis 对 MapReduce 中的数据进行连接。

1. 创建 Java 方法从当前的工作文件夹读取文件 nobots_ip_country_tsv.txt，使用 Jedis 客户端载入相关内容至 Redis：

```
private void loadRedis(String ipCountryTable) throws IOException {
FileReader freader = new FileReader(ipCountryTable);
    BufferedReader breader = new BufferedReader(freader);
    jedis = new Jedis("localhost");
    jedis.select(0);
    jedis.flushDB();
    String line = breader.readLine();
    while(line != null) {
        String[] tokens = line.split("\t");
        String ip = tokens[0];
        String country = tokens[1];
        jedis.set(ip, country);
        line = breader.readLine();
    }
    System.err.println("db size = " + jedis.dbSize());
}
```

2. 设置只有 map 的 MapReduce 作业。下面的代码是创建只有 map 的 MapReduce 作业对应的类的最终版本。包含在步骤一中创建的方法 loadRedis()：

```
public class MapSideJoinRedis extends Configured implements Tool {
```

```java
  private Jedis jedis = null;

  private void loadRedis(String ipCountryTable) throws
    IOException {

    FileReader freader = new FileReader(ipCountryTable);
    BufferedReader breader = new BufferedReader(freader);
    jedis = new Jedis("localhost");
    jedis.select(0);
    jedis.flushDB();
    String line = breader.readLine();
    while(line != null) {
      String[] tokens = line.split("\t");
      String ip = tokens[0];
      String country = tokens[1];
      jedis.set(ip, country);
      line = breader.readLine();
    }
    System.err.println("db size = " + jedis.dbSize());
  }

  public int run(String[] args) throws Exception {

    Path inputPath = new Path(args[0]);
    String ipCountryTable = args[1];
    Path outputPath = new Path(args[2]);

    loadRedis(ipCountryTable);

    Configuration conf = getConf();
    Job weblogJob = new Job(conf);
    weblogJob.setJobName("MapSideJoinRedis");
    weblogJob.setNumReduceTasks(0);
    weblogJob.setJarByClass(getClass());
    weblogJob.setMapperClass(WeblogMapper.class);
    weblogJob.setMapOutputKeyClass(Text.class);
    weblogJob.setMapOutputValueClass(Text.class);
    weblogJob.setOutputKeyClass(Text.class);
    weblogJob.setOutputValueClass(Text.class);
    weblogJob.setInputFormatClass(TextInputFormat.class);
    weblogJob.setOutputFormatClass(TextOutputFormat.class);
    FileInputFormat.setInputPaths(weblogJob, inputPath);
    FileOutputFormat.setOutputPath(weblogJob, outputPath);

    if(weblogJob.waitForCompletion(true)) {
      return 0;
    }
    return 1;
  }

  public static void main(String[] args) throws Exception {
      int returnCode = ToolRunner.run(new
```

```
            MapSideJoinRedis(), args);
        System.exit(returnCode);
    }
}
```

3. 创建 mapper 连接数据集 apache_nobots_tsv.txt 与加载至 Redis 的数据集 nobots_ip_country_tsv.txt：

```
public class WeblogMapper extends Mapper<Object, Text, Text, Text>
{
    private Map<String, String> ipCountryMap = new
      HashMap<String, String>();
    private Jedis jedis = null;
    private Text outputKey = new Text();
    private Text outputValue = new Text();

    private String getCountry(String ip) {
        String country = ipCountryMap.get(ip);
        if (country == null) {
            if (jedis == null) {
                jedis = new Jedis("localhost");
                jedis.select(0);
            }
            country = jedis.get(ip);
            ipCountryMap.put(ip, country);
        }
        return country;
    }

    @Override
    protected void map(Object key, Text value, Context
      context) throws IOException, InterruptedException {
        String row = value.toString();
        String[] tokens = row.split("\t");
        String ip = tokens[0];
        String country = getCountry(ip);
        outputKey.set(country);
        outputValue.set(row);
        context.write(outputKey, outputValue);
    }
}
```

4. 最后，启动 MapReduce 作业：

```
$ hadoop jar AdvJoinChapter5-1.0-SNAPSHOT.jar com.packt.ch5.advjoin.redis.
MapSideJoinRedis   /user/hadoop/apache_nobots_tsv.txt./nobots_ip_country_tsv.
txt /user/hadoop/data_jnd
```

工作原理

在第一步和第二步，创建一个类用于设置只有 map 的作业。除了 loadRedis() 方法之外，这个类与之前章节中创建只有 map 的作业的类相似。

　　loadRedis() 方法首先使用 Jedis 构造器连接本地的 Redis 实例。接下来，使用 select() 方法选择我们需要使用的数据库。单个 Redis 实例可以包含多个数据库，使用数字索引对其标识。一旦连接上想要的数据库后，调用 flushDB() 方法，将会删除当前数据库的所有存储内容。最后，从当前工作文件夹读取文件 nobots_ip_country_tsv.txt，并使用 set() 方法载入 Redis 实例的键值对 ip/country。

更多参考

　　本节使用一个非常简单的字符串数据结构存储 ip/country 键值对。Redis 支持其他多种数据结构包括散列表、列表以及有序集合。此外，Redis 支持事务、发布/订阅机制。如果要深入了解这些功能请访问 Redis 网站 http://redis.io/。

第 6 章

大数据分析

本章我们将介绍:
- 使用 MapReduce 和 Combiner 统计网络日志数据集中的独立 IP 数
- 运用 Hive 日期 UDF 对地理事件数据集中的时间日期进行转换与排序
- 使用 Hive 创建基于地理事件数据的每月死亡报告
- 实现 Hive 用户自定义 UDF 用于确认地理事件数据的来源可靠性
- 使用 Hive 的 map/reduce 操作以及 Python 标记最长的无暴力发生的时间区间
- 使用 Pig 计算 Audioscrobbler 数据集中艺术家之间的余弦相似度
- 使用 Pig 以及 datafu 剔除 Audioscrobbler 数据集中的离群值

6.1 介绍

学会运用 Apache Hive、Pig 以及 MapReduce 解决你所要面对的特定的问题,有时候是具有挑战性的。这章将介绍一些有关大数据分析的问题以及解决这些问题的方法。你将会发现,需要解决的问题并非难以置信地复杂,只是在处理大规模数据的时候,需要一些特殊的解决方法。即使本章采样的数据集较小,这些代码对于大数据集同样非常适用。

本章对需要分析的问题做了精心设计,以展示各种工具的强大特性。你会发现在解决问题的时候,这些特性和功能十分有用。

6.2 使用 MapReduce 和 Combiner 统计网络日志数据集中的独立 IP 数

本节将指导你创建一个 MapReduce 程序，统计在网络日志数据集中的独立 IP 数。我们将展示 combiner 特性，它能用于优化减少 map 和 reduce 之间传输的数据量。代码的实现适用于通用的形式，能对任何制表符分割的数据集中不同的值进行统计。

准备工作

本节假设你已经对 Hadoop 0.20 MapReduce API 已经有基本的了解。你需要访问本书提供的数据集 `weblog_entries`，并将其存储于 HDFS 路径 `/input/weblog` 中。

使用 Hadoop 0.20 提供的新的 MapReduce API，在伪分布式或完全分布式集群运行 MapReduce 作业。

你同样需要打包这些代码至一个 JAR 文件并在 shell 中使用 Hadoop JAR Launcher 运行这个 JAR 文件。运行这个例子只需要编译核心的 Hadoop 库。

操作步骤

执行下面的步骤，使用 MapReduce 统计独立的 IP 数。

1. 打开一个你常用的文本编辑器/IDE，最好具有 Java 语法高亮功能。

2. 在 JAR 文件合适的源软件包中，创建一个 `DistinctCounterJob.java` 类。

3. 为了能正确提交作业，下面的代码是一个 Tool 的实现：

```java
import org.apache.hadoop.conf.Configuration;
import org.apache.hadoop.fs.Path;
import org.apache.hadoop.io.IntWritable;
import org.apache.hadoop.io.LongWritable;
import org.apache.hadoop.io.Text;
import org.apache.hadoop.mapreduce.Job;
import org.apache.hadoop.mapreduce.Mapper;
import org.apache.hadoop.mapreduce.Reducer;
import org.apache.hadoop.mapreduce.lib.input.FileInputFormat;
import org.apache.hadoop.mapreduce.lib.input.TextInputFormat;
import org.apache.hadoop.mapreduce.lib.output.FileOutputFormat;
import org.apache.hadoop.mapreduce.lib.output.TextOutputFormat;
import org.apache.hadoop.util.Tool;
import org.apache.hadoop.util.ToolRunner;
import java.io.IOException;
import java.util.regex.Pattern;

public class DistinctCounterJob implements Tool {
```

```
    private Configuration conf;
    public static final String NAME = "distinct_counter";
    public static final String COL_POS = "col_pos";

    public static void main(String[] args) throws Exception {
        ToolRunner.run(new Configuration(), new DistinctCounterJob(), args);
    }
```

4. `run()`方法设置了输入/输出格式，mapper 类的、combiner 类的以及键值类的配置：

```
    public int run(String[] args) throws Exception {
        if(args.length != 3) {
            System.err.println("Usage: distinct_counter <input> <output> <element_position>");
            System.exit(1);
        }
        conf.setInt(COL_POS, Integer.parseInt(args[2]));

        Job job = new Job(conf, "Count distinct elements at position");
        job.setInputFormatClass(TextInputFormat.class);
        job.setOutputFormatClass(TextOutputFormat.class);

        job.setMapperClass(DistinctMapper.class);
        job.setReducerClass(DistinctReducer.class);
        job.setCombinerClass(DistinctReducer.class);

        job.setMapOutputKeyClass(Text.class);
        job.setMapOutputValueClass(IntWritable.class);
        job.setJarByClass(DistinctCounterJob.class);

        FileInputFormat.addInputPath(job, new Path(args[0]));
        FileOutputFormat.setOutputPath(job, new Path(args[1]));

        return job.waitForCompletion(true) ? 1 : 0;
    }

    public void setConf(Configuration conf) {
        this.conf = conf;
    }

    public Configuration getConf() {
        return conf;
    }
}
```

5. 下面代码实现的 `map()` 方法扩展了 `mapreduce.Mapper` 类：

```
    public static class DistinctMapper
            extends Mapper<LongWritable, Text, Text, IntWritable> {

        private static int col_pos;
        private static final Pattern pattern = Pattern.compile("\\t");
        private Text outKey = new Text();
        private static final IntWritable outValue = new IntWritable(1);
```

```
        @Override
        protected void setup(Context context
        ) throws IOException, InterruptedException {
            col_pos = context.getConfiguration().getInt(DistinctCounterJob.
COL_POS, 0);
        }

        @Override
        protected void map(LongWritable key, Text value, Context context)
throws IOException, InterruptedException {
            String field = pattern.split(value.toString())[col_pos];
            outKey.set(field);
            context.write(outKey, outValue);
        }
    }
```

6. 下面代码实现的 reduce() 方法扩展了 mapreduce.Reducer 类：

```
    public static class DistinctReducer
            extends Reducer<Text, IntWritable, Text, IntWritable> {

        private IntWritable count = new IntWritable();

        @Override
        protected void reduce(Text key, Iterable<IntWritable> values, Context
context
        ) throws IOException, InterruptedException {
            int total = 0;
            for(IntWritable value: values) {
                total += value.get();
            }
            count.set(total);
            context.write(key, count);
        }
    }
```

7. 下面的命令展示了对网络日志数据集的用法示例，其中 IP 对应的列的位置为 4：

```
hadoop jar myJobs.jar distinct_counter /input/weblog/ /output/ weblog_
distinct_counter 4
```

工作原理

首先，创建了 DistinctCounterJob，实现了用于远程提交的 Tool 接口。静态常量 NAME 在 Hadoop Driver 类中被非显式使用，用于支持不同作业对同一 JAR 文件的启动。静态常量 COL_POS 从命令行 <element_position> 的第三个参数获得初始值。这个值在作业配置阶段设置，需要与希望统计的唯一实体值所在的列的位置一致。对于网络日志数据的 IP 列，设定为 4。

由于需要读写文本，需要分别指定类 TextInputFormat 以及类 TextOutputFormat。作为实现类 DistinctMapper 和 DistinctReducer 的 Mapper 类以及 Reducer 类。同时

将 `DistinctReducer` 作为 combiner 类。

同样,调用 `setJarByClass()` 也是非常重要的,因为 `TaskTrackers` 需要能正确地解压 JAR 文件并找到其中的 `Mapper` 和 `Reducer` 类。该作业使用类 `FileInputFormat` 和类 `FileOutputFormat` 的静态辅助方法,分别设置输入输出路径。现在已经设置并准备提交该作业。

`Mapper` 类的实现中,设置如下一些成员变量。

- `col_pos`:这是在配置中需要设定的初始化值。允许用户设定并修改哪一行是执行解析并统计不同值操作的对象。
- `pattern`:对于每一行,定义列的分隔符,这里是制表符。
- `outKey`:这是类的成员变量,记录输出的值。这样是为了避免必须为每行的输出创建一个新的实例。
- `outValue`:这是一个整数,表示给定的键出现了一次。这里与例子 WordCount 类似。

`map()` 分割每行输入的值,抽取位于 `col_pos` 的字符串。重新将位于这个行相应位置的字符串作为 `outKey` 的值。对于本例而言,会是每行的 IP 值。mapper 将重新设置的 `outKey` 变量与标记为给定 IP 地址出现一次的 `outValue` 一同输出。

没有 combiner 的支持,将会出现 reducer 需要对一个只包含数字 1 的集合进行计数。

下面的例子是不包含 combiner 的 reducer{key,value:[]}的情况:

{10.10.1.1, [1,1,1,1,1,1]} =六次出现 IP"10.10.1.1"

`reduce ()` 方法的实现将对出现的整数进行正确地求和,但这里不仅仅将整数的值限定为数字 1。可以使用 combiner 处理中间的键值对作为每个 mapper 的输出,改善在 shuffle 阶段的数据吞吐量。由于 combiner 是应用于每个本地 map 的输出,可以看到需要传输的大量中间 key/value 数据被大大减少,从而性能得到提升。

为了替换{10.10.1.1, [1,1,1,1,1,1]},combiner 累加每个 IP 的出现次数,替换这个键对应的中间结果为{10.10.1.1, [6]}。reducer 能对合并之后的值进行正确地求和。这是基于加法运算满足交换律与结合律。或者可以这么说:

- **交换律**:改变加法运算对象的顺序不影响最后的结果。比如,1 + 2 + 3 = 3 + 1 + 2。
- **结合律**:应用加法操作的顺序不影响最后的结果。比如,(1 + 2) + 3 = 1 + (2 + 3)。

为了计算独立 IP 的出现次数,对于 map 阶段的输出,可以使用与 reducer 一样的代码作为 combiner。

对于我们的问题，不包含 combiner 的两个独立运行的 map 任务的正常输出结果如下所示，其中{key: value[]}表示键值集合的中间结果：

- Map Task A = {10.10.1.1, [1,1,1]} = 出现 3 次
- Map Task B = {10.10.1.1, [1,1,1,1,1,1]} =出现 6 次

没有 combiner 的帮助，在 shuffle 阶段中表示为如下形式的键值集合将在一个 reducer 中被合并：

- {10.10.1.1, [1,1,1,1,1,1,1,1,1]} =总共出现 9 次

现在让我们重新回顾下，对同样的输出结果，使用 Combiner：

- Map Task A = {10.10.1.1, [1,1,1]} =出现 3 次

Combiner = {10.10,1,1, [3] }=仍然表示出现 3 次，但对这个 map 进行合并

- Map Task B = {10.10.1.1, [1,1,1,1,1,1]} =出现 3 次

Combiner = {10.10,1,1, [6]} =仍然表示出现 6 次

接着，reducer 将会得到如下的键值集合：

{10.10,1,1, [3,6]}=总共出现 9 次

我们得到对于不同 IP 地址的总的统计数，同时使用 combiner，通过提前归并每个 mapper 的键值中间结果，降低了在 MapReduce 的 shuffle 阶段的网络 I/O 的总和。

更多参考

对新手而言，combiner 的使用有时候会带来一些疑惑。下面是一些有用的小技巧。

combiner 不总是和 Reducer 相同

上节的例子以及 WordCount 的例子中的 `Combiner` 类都与 `Reducer` 类使用相同的实现类进行初始化。不是 API 强制这样规定的，而是由于很多种分布式聚合操作本身导致的，比如：`sum()`、`min()`以及 `max()`。一个简单的例子是对于 `Reducer` 类的 `min()`操作需要依据一些特定的输出格式。而对于 `Combiner` 类的 `min()`操作形式上会有轻微地不同，不需要关心特定的输出格式。

combiner 不保证一定运行

框架是否在执行过程中调用 combiner 依赖于每个 map 输出的中间分片文件大小，因此不保证每个中间键都运行对应的 combiner。作业结果的正确性也不依赖于 combiner，

combiner 只用于优化作业的执行过程。

当 MapReduce 试图调用 combiner 合并中间结果的时候，可以通过设置配置项 `min.num.spills.for.combine` 控制分片文件的阈值。

6.3 运用 Hive 日期 UDF 对地理事件数据集中的时间日期进行转换与排序

本节将展示如何高效地使用 Hive 提供的时间 UDF，列出最近的 20 个事件以及事件发生时间距当前系统时间的天数。

准备工作

确保已在你的客户端安装能访问伪分布式或全分布式的 Hadoop 集群的 Apache Hive 0.7.1，并保证当前用户的环境路径正确。

本节需要将数据集 Nigera_ACLED_cleaned.tsv 加载为 Hive 表 acled_nigeria_cleaned，且下面的字段分别对应正确的数据类型。

在 Hive 客户端输入下面的命令：

```
describe acled_nigeria_cleaned
```

将返回如下结果：

```
OK
Loc      string
event_date  string
event_type  string
actor    string
latitude  double
longitude double
source    string
fatalities  int
```

操作步骤

执行下面的步骤，利用 Hive 的 UDF 进行排序与数据转换。

1. 打开一个常用的文本编辑器，最好具有 SQL 语法高亮显示功能。

2. 添加如下的内联创建与转换语句：

```
SELECT event_type,event_date,days_since FROM (
    SELECT event_type,event_date,
           datediff(to_date(from_unixtime(unix_timestamp())),
```

```
                to_date(from_unixtime(
                    unix_timestamp(event_date,
                    'yyyy-MM-dd')))) AS days_since
    FROM acled_nigeria_cleaned) date_differences
    ORDER BY event_date DESC LIMIT 20;
```

3. 在当前文件夹保存上述文件为：`top_20_recent_events.sql`。

4. 在系统 shell 中，从添加`-f`选项的 Hive 客户端运行该脚本，将会首先在命令行中看见如下的 5 行输出：

```
OK
Battle-No change of territory   2011-12-31   190
Violence against civilians      2011-12-27   194
Violence against civilians      2011-12-25   196
Violence against civilians      2011-12-25   196
Violence against civilians      2011-12-25   196
```

工作原理

我们首先分析嵌套的 `SELECT` 子查询。我们选取了来自 Hive 表 `acled_nigeria_cleaned` 的 3 个字段：`event_type`、`event_date` 以及调用 `datediff()` 得到的结果，这个 UDF 会将结束时间和开始时间作为参数。这两个时间都是 `yyyy-MM-dd` 的格式。`datediff()` 的第一个参数为结束时间，这里我们表示为当前系统时间。调用不包含参数的 `unix_timestamp()` 函数会以毫秒的形式返回当前系统时间。将返回的值作为函数 `from_unixtimestamp()` 的参数，当前的系统时间则会表达为默认 Java 1.6 的时间戳格式(yyyy-MM-dd HH:mm:ss)。由于我们只关注日期部分，因此调用 `to_date()` 去掉输出的 HH:mm:ss 部分。此时，当前时间表示为 `yyyy-MM-dd` 的形式。

`datediff()` 的第二个参数为开始时间，这里是查询中的 `event_date` 字段。除了在调用 `unix_timestamp()` 的时候需要告知函数参数使用的是 `SimpleDateFormat` 格式 `yyyy-MM-dd` 之外，这里一系列的函数调用与之前方法的参数是大致相同的。现在，参数 `start_date` 与 `end_date` 都是 `yyyy-MM-dd` 格式，可以对给定的列执行 `datediff()` 操作。然后，将 `datediff()` 的输出结果对应的列命名别名为 `days_since`。

为了获得逆序的时间列表，外层的 `SELECT` 语句选取了每行的三列值，并按照 `event_date` 逆序排列输出结果。这里我们任意设定，将输出的结果限定为前 20 行。

最终的结果是至今最近的 20 个事件以及对应事件发生距今的天数。

更多参考

时间 UDF 能极大的有助于执行日期字符串的比较。下面是一些额外的知识。

日期格式字符串遵循 Java 的 SimpleDateFormat 标准

阅读关于 `SimpleDateFormat` 的 Java 文档，学会如何使用日期转换 UDF 处理你自定义的日期字符串。

默认的日期与时间格式

- 很多的 UDF 操作假设按照默认的日期格式执行。
- 对于只需要日期的 UDF，对应列的格式必须是 yyyy-MM-dd。
- 对于既需要日期又需要时间的 UDF，对应列的格式必须是 yyyy-MM-dd HH:mm:ss。

延伸阅读

- 使用 Hive 创建基于地理事件数据的每月死亡报告（6.4 节）。

6.4 使用 Hive 创建基于地理事件数据的每月死亡报告

本节将展示一个使用 Hive 的简单分析，统计出现在数据集中每个月的死亡人数并将其打印至控制台。

准备工作

确保已在你的客户端安装能访问伪分布式或全分布式的 Hadoop 集群的 Apache Hive 0.7.1，并保证当前用户的环境路径正确。

本节需要将数据集 `Nigera_ACLED_cleaned.tsv` 加载为 Hive 表 `acled_nigeria_cleaned`，且下面的字段分别对应正确的数据类型。

在 Hive 客户端输入下面的命令：

```
describe acled_nigeria_cleaned
```

将返回如下结果：

```
OK
loc        string
event_date    string
event_type    string
actor      string
latitude   double
longitude  double
source     string
fatalities    int
```

操作步骤

执行下面的步骤,利用 Hive 生成相关报告。

1. 打开一个常用的文本编辑器,最好具有 SQL 语法高亮显示功能。

2. 添加如下的内联创建与转换语句:

```
SELECT from_unixtime(unix_timestamp(event_date, 'yyyy-MM-dd'),
'yyyy-MMM'),
       COALESCE(CAST(sum(fatalities) AS STRING), 'Unknown')
       FROM acled_nigeria_cleaned
       GROUP BY from_unixtime(unix_timestamp(event_date, 'yyyy-MM-dd'),
'yyyy-MMM');
```

3. 在当前文件夹保存上述文件为:monthly_violence_totals.sql。

4. 在系统 shell 中,从添加-f 选项的 Hive 客户端运行该脚本,将会首先在命令行中出现如下的三行输出。注意:输出的结果按照字典序排序而非按照时间排序。

```
OK
1997-Apr    115
1997-Aug    4
1997-Dec    26
```

工作原理

SELECT 语句使用 unix_timestamp() 和 from_unixtime() 方法,将每行记录的 event_date 字段格式化为年份-月份连接的形式。在 GROUP BY 表达式中使用 sum() 对死亡人数进行统计。

coalesce() 方法返回第一个非空的参数。这里将给定年月的死亡人数总和作为第一个参数,并将其转化为一个字符串。如果此值由于某些原因为 NULL,将返回常量 Unkown;否则将返回年月时间组合对应的死亡总人数并表示为一个字符串。将这些结果通过 stdout 输出至控制台。

更多参考

关于本节中的一些代码,有如下的一些额外的小技巧。

coalesce()能将变长参数作为参数

Hive 的官方文档提到,coalesce() 支持一个或多个参数。第一个非空的参数将会被返回。这对于评估一个给定的列的几个不同的表达式(在决定选择哪个是正确的一个之前)是很有用的。

如果没有非空的参数,coalesce() 会返回 NULL。如果其他参数都为 NULL,返回一个文本类型的值也是不少见的情况。

时间重格式化代码模板

对原始数据的日期字段进行重新格式化是非常常见的操作。正确地使用 `from_unixtime()` 和 `unix_timestamp()` 能使你的工作更加轻松。

记下这个在 Hive 中通用的对简单日期格式进行转换的代码模板：

```
from_unixtime(unix_timestamp(<col>,<in-format>),<out-format>);
```

延伸阅读

- 运用 Hive 日期 UDF 对地理事件数据集中的时间日期进行转换与排序（6.3 节）。

6.5 实现 Hive 用户自定义 UDF 用于确认地理事件数据的来源可靠性

在 Hive 中，有许多的操作在不同的数据源和表中反复执行。对于这种情况，创建用户自定义函数（UDF）是非常有意义的。你可以使用 Java 对任何输入 Writable 的字段编写子程序，在 Hive 中任何需要的时候调用。本节将执行一个非常简单的 UDF 用来对数据源的可靠性进行 yes/no 的判断。

准备工作

确保已在你的客户端安装能访问伪分布式或全分布式 Hadoop 集群的 Apache Hive 0.7.1，并保证当前用户的环境路径正确。

本节需要将数据集 `Nigera_ACLED_cleaned.tsv` 加载为 Hive 表 `acled_nigeria_cleaned`，且下面的字段分别对应正确的数据类型。

在 Hive 客户端输入下面的命令：

```
describe acled_nigeria_cleaned
```

将返回如下结果：

```
OK
loc         string
event_date  string
event_type  string
actor       string
latitude    double
longitude   double
source      string
fatalities  int
```

此外，需要将本节下面的代码放置在一个 JAR 文件的源软件包中。本节使用 `<myUDFs.jar>`

指向用户自定义的 JAR 文件，<fully_qualified_path_to_TrustSourceUDF>指向保存你自己编写的类的 Java 包。例如，模式的一个正确的完整路径为 java.util.regex.Pattern。

此外，由于需要 Hadoop 核心库，该工程需要在编译的 classpath 中包括 hive-exec 以及 hive-common JAR 依赖。

操作步骤

执行下面的步骤，实现一个用户自定义的 Hive UDF。

1. 打开一个你常用的文本编辑器/IDE，最好拥有 Java 语法高亮功能。

2. 创建 TrustSourceUDF.java 需要的源软件包。对应的类应该保存在同样的包中，即<fully_qualified_path>.TrustSourceUDF.class。

3. 输入下面的源代码实现 TrustSourceUDF 类：

```java
import org.apache.hadoop.hive.ql.exec.UDF;
import org.apache.hadoop.io.Text;
import java.lang.String;
import java.util.HashSet;
import java.util.Set;

public class TrustSourceUDF extends UDF {

    private static Set<String> untrustworthySources = new HashSet<String>();
    private Text result = new Text();

    static {
     untrustworthySources.add("");
     untrustworthySources.add("\"\"\"
     http://www.afriquenligne.fr/3-soldiers\"");
     untrustworthySources.add("Africa News Service");
     untrustworthySources.add("Asharq Alawsat");
     untrustworthySources.add("News Agency of Nigeria (NAN)");
     untrustworthySources.add("This Day (Nigeria)");
    }

    @Override
    public Text evaluate(Text source) {
        if(untrustworthySources.contains(source.toString())) {
            result.set("no");
        } else {
            result.set("yes");
        }
        return result;
    }
}
```

大数据分析

4. 创建 JAR 文件 <myUDFs.jar>，并通过 Hive 客户端测试自定义的 UDF。通过命令行 shell 打开一个 Hive 客户端。Hive 已经在当前用户的环境路径中设置好。使用下面的命令，调用 Hive shell：

```
hive
```

5. 在 Hive 的会话路径中添加 JAR 文件：

```
add jar /path/to/<myUDFs.jar>;
```

如果出现如下的信息表明 JAR 已经被加入至 classpath 以及分布式缓存，表明此操作已运行成功：

```
Added /path/to/<myUDFs.jar> to class path
Added resource: /path/to/<myUDFs.jar>
```

6. 对 JAR 中指定的源软件包中的 `TrustSourceUDF` 创建函数定义别名 `trust_source`：

```
create temporary function trust_source as '<fully_qualified_path_to_TrustSourceUDF>';
```

shell 中会提示命令已经执行成功。如果出现如下的错误信息，通常表明在类路径下无法找到你创建的类：

```
FAILED: Execution Error, return code 1 from org.apache.hadoop.hive.ql.exec.FunctionTask
```

7. 通过下面的查询测试该函数。将在命令行中出现大量的 **yes**，同时也会出现少量的 **no**。

```
select trust_source(source) from acled_nigeria_cleaned;
```

工作原理

类 `TrustSourceUDF` 继承了 UDF。不需要实现任何的方法，然而，为了在 Hive 中将类的方法作为一个 UDF，需要重载方法 `evaluate()`。可以重载多个不同参数的 `evaluate()` 方法。这里，我们只需要将 source 的值进行检查判断。

类初始化的时候，创建了类 `java.util.Set` 的一个静态实例 `untrustworthySources`。在静态初始化代码块中，我们根据源的名称将其标记为不可靠。

 这些数据源是否可靠的判定是完全随意的，因此这里的结果不适合于本节之外的其他地方

我们将数据源为空也作为不可靠的。

当函数被调用的时候，单个 `Text` 实例会与已标记为不可靠的数据源作对比。返回 yes 或 no 依赖于给定的数据源是否出现在不可靠的数据源集合中。创建的私有变量 Text 用于存储每次函数被调用的返回值。

一旦包含这个类的 JAR 文件被添加至 classpath 中，临时的函数定义被创建，我们可以

在很多不同的查询中使用这个 UDF。

> **更多参考**

在 Hive 中，用户自定义函数是非常有用的一个特性。下面列出了与之相关的更多内容：

确认已存在的 UDF

Hive 文档对语言内置的 UDF 提供了很好的解释。可以从 https://cwiki.apache.org/confluence/display/Hive/LanguageManual+UDF#LanguageManualUDF-BuiltinAggregateFunctions%28UDAF%29 获得相应的文档。

在你当前的 Hive 版本下，可以在 Hive shell 中输入下面的命令获得哪些函数是可以使用的：
```
show functions;
```

一旦定位到感兴趣的函数，可以通过 Hive wiki 或者直接在 Hive shell 中执行下面命令获得更多信息：
```
describe function <func>;
```

用户自定义表与聚合函数

Hive UDF 不需要一对一的输入输出交互。API 允许从一个输入生成很多的输出（GenericUDTF），同时用户聚合函数将一个输入行的列表输出为一个单值（UDAF）。

在当前环境下导出 `HIVE_AUX_JARS_PATH`

对于测试与调试，动态地添加 JAR 文件到 classpath 中是有用的，然而经常添加一些频繁使用的库文件是非常繁琐的。Hive 命令行解释器会自动地检查当前用户执行环境的 `HIVE_AUX_JARS_PATH` 是否存在。使用该环境变量，可以设置需要添加的 JAR 的路径，这样该客户机的每个新的 Hive 会话都会将 JAR 文件加载至 classpath 中。

> **延伸阅读**
>
> ❑ 运用 Hive 日期 UDF 对地理事件数据集中的时间日期进行转换与排序（6.3 节）；
> ❑ 使用 Hive 创建基于地理事件数据的每月死亡报告（6.4 节）。

6.6 使用 Hive 的 map/reduce 操作以及 Python 标记最长的无暴力发生的时间区间

Hive 查询语句提供了控制 MapReduce 数据流、在各个阶段注入自定义 map/reduce 脚本的方法。恰当地运用该方法能更简洁地表达 MapReduce 程序。

本节将完整地展示如何在 Hive 中使用不同的操作编写用户自定义 MapReduce 控制流。这个分析关注发生在各个地点的事件的最长时间区间，从而获得该地点暴力发生的频率。

准备工作

确保已在你的客户端安装能访问伪分布式 Hadoop 集群或全分布式 Hadoop 集群的 Apache Hive 0.7.1，并保证当前用户的环境路径正确。

集群的每个节点需要安装 Python 2.7 或更高的版本，并且 Hadoop 用户的环境路径可用。本节脚本假设 Python 安装至 /usr/bin/env。如果和你的安装不一致，请修改相关脚本。

本节需要将数据集 Nigera_ACLED_cleaned.tsv 加载为 Hive 表 acled_nigeria_cleaned，且下面的字段分别对应正确的数据类型。

在 Hive 客户端输入下面的命令：

```
describe acled_nigeria_cleaned
```

将返回如下结果：

```
OK
loc         string
event_date  string
event_type  string
actor       string
latitude    double
longitude   double
source      string
fatalities  int
```

操作步骤

执行下面的步骤，使用 Hive 标记最长的非暴力时间区间。

1. 打开一个你常用的文本编辑器，最好拥有 SQL 以及 Python 语法高亮功能。

2. 添加下列内部创建与转换语句：

```
SET mapred.child.java.opts=-Xmx512M;

DROP TABLE IF EXISTS longest_event_delta_per_loc;
CREATE TABLE longest_event_delta_per_loc (
    loc STRING,
    start_date STRING,
    end_date STRING,
    days INT
);
```

```
ADD FILE calc_longest_nonviolent_period.py;
FROM (
        SELECT loc, event_date, event_type
        FROM acled_nigeria_cleaned
        DISTRIBUTE BY loc SORT BY loc, event_date
    ) mapout
INSERT OVERWRITE TABLE longest_event_delta_per_loc
REDUCE mapout.loc, mapout.event_date, mapout.event_type
USING 'python calc_longest_nonviolent_period.py'
AS loc, start_date, end_date, days;
```

3. 在当前工作路径保存文件为 `longest_nonviolent_periods_per_location.sql`。

4. 在文本编辑器创建文件 `calc_longest_nonviolent_period.py`，并保存在与文件 `longest_nonviolent_periods_per_location.sql` 相同的文件夹下。

5. 添加 Python 语句。由于 Python 是缩进敏感的，当你复制粘贴下面的代码时要十分小心：

```python
#!/usr/bin/python
import sys
from datetime import datetime, timedelta

current_loc = "START_OF_APP"
(prev_date, start_date, end_date, start_time_obj, end_time_obj,
current_diff)=('', '', '', None, None, timedelta.min)
for line in sys.stdin:
  (loc,event_date,event_type) = line.strip('\n').split('\t')
  if loc != current_loc and current_loc != "START_OF_APP":
    if end_date != '':
      print '\t'.join([current_loc,start_date,event_date, str(current_diff.days)])
          (prev_date, start_date, end_date, start_time_obj, end_time_obj,current_diff)=('', '', '', None, None, timedelta.min)
    end_time_obj = datetime.strptime(event_date,'%Y-%m-%d')
  current_loc = loc
  if start_time_obj is not None: # implies > 2 events
    diff = end_time_obj - start_time_obj
    if diff > current_diff:
      current_diff = diff # set the current max time delta
      start_date = prev_date
      end_date = event_date
  prev_date = event_date
  start_time_obj = end_time_obj
```

6. 从系统 shell 执行添加 `-f` 选项的脚本至 Hive 客户端：

```
hive -f longest_nonviolent_periods_per_location.sql
```

7. 在 Hive shell 中直接进行查询。相关的行记录未按照特定的顺序输出至命令行。

```
hive -e "select * from longest_event_delta_per_loc;"
```

工作原理

让我们首先分析创建的 Hive 脚本。第一行简单设定了执行 JVM 的堆大小。可以根据你的集群设置合适的值。对于数据集 ACLED Nigeria，512 MB 是足够的了。

然后，创建了输出的表的定义，并删除已存在的表 `longest_event_delta_per_loc`。这个表包含 4 个字段：`loc` 表示位置，`start_date` 表示值 `event_date` 的下限，`end_date` 表示值 `event_date` 的上限，`day` 表示事件之间间隔的天数。

接下来，添加文件 `calc_longest_nonviolent_period.py` 至分布式缓存，供不同的 reducer JVM 使用。这将作为 reduce 脚本运行，但首先我们需要组织 map 的输出。最内层的 `SELECT` 语句抓取 Hive 表中 `acled_nigeria_cleaned` 的字段 `loc`、`event_date` 以及 `event_type`。语句 `DISTRIBUTE BY loc` 告知 Hive，所有同样 `loc` 值的行发送至同一个 reducer 中。`SORT BY loc, event_date` 告知 Hive 对每个发送至 reducer 的数据，根据 `loc` 与 `event_loc` 的组合进行排序。此时，同一个 reducer 能以 `event_date` 排序，处理与每个地点相关的行记录。

将这个 `SELECT` 语句的输出取别名为 `mapout` 并使用简写 `REDUCE` 操作处理来自 `mapout` 的每行记录。`USING` 子句从 `stdin` 读取每行记录运行自定义 Python 脚本。`AS` 操作符将脚本从 `stdout` 输出的分割字段映射至接收表的相关字段。

Python 脚本 `calc_longest_nonviolent_period.py` 将在 reduce 阶段为每个地点计算最长的两个事件之间的时间间隔。由于我们保证了同一个 `loc` 值的记录都在同一个 reducer 中，并且这些关于地点的记录都按照时间排序。现在，我们能了解 Python 脚本是如何工作的了。

在 Python 脚本 `calc_longest_nonviolent_period.py` 中，开始的 `#!/usr/bin/python` 提示 shell 如何执行该脚本。为了使用 `stdin`、`stdout` 操作，我们需要导入 `sys`。我们同样需要使用来自 `datetime` 包的 `datetime`、`timedelta` 类。

脚本的相关操作非常程序化，读起来会有一些难理解。首先声明了 `current_loc` 并初始化值为 `START_OF_APP` 作为输出条件的表示。接下来，设置了一些变量用于保存 `for` 循环中基于各个位置的占位符。

- ❏ `prv_date`：保存 `loc` 值最近出现的 `event_date`。如果是应用的开始或为新的位置，则该值为空。

- ❑ `start_date`：保存对于位置 `loc`，已出现的最长事件时间区间的下限值。
- ❑ `end_date`:保存对于位置 `current_loc`，已出现的最长事件时间区间的上限值。
- ❑ `start_time_obj`：保存最近迭代的 `datetime` 对象。如果是应用的开始或为新的位置，则该值为 `None`。
- ❑ `end_time_obj`：保存当前 `event_date` 的 `datetime` 对象，如果是应用的开始或为新的位置，则该值为 `None`。
- ❑ `current_diff`：保存对于 `current_loc`，当前的最长事件时间区间值，如果是应用的开始或为新的位置，则该值是一个在 time delta 中定义的最小的时间差值。

`for` 循环从 `stdin` 读取已经根据 `loc` 和 `event_date` 组合排序的行记录。对新的行按照制表符进行分隔，解析每行记录至相关的列变量中。

首先，第一个条件是跳过 `current_loc` 等于 `START_OF_APP` 的条目。当处理 reducer 上所有位置的第一行数据时，不输出任何结果。当 `loc` 的值不等于 `current_loc` 的值且不位于应用开始的时刻，这表明已经处理完所有 `current_loc` 的记录，此时可以最终输出该地点最长的事件时间间隔。如果 `end_date` 仍然是一个空串，表明当前地点只包含一个事件。在这种情况下，对于该地点不输出任何结果。最后，重置上面提及的 6 个参数，处理下一个位置的记录。

接下来，设置 `current_loc` 的值为当前处理的 `loc` 值，从而避免下一次迭代不满足上述条件的条目。设置 `end_time_obj` 为当前行的 `event_date` 的值。如果 `start_time_obj` 赋值为 `None`，表明这是当前位置的第一行记录，不能进行时间间隔的比较。无论 `start_time_obj` 是否设置为 `None`，在循环的结尾处设置 `prev_date` 为 `event_date`，设置 `start_time_obj` 为当前遍历的 `end_time_obj`。在下一次迭代的时候，`start_time_obj` 将会保存之前记录的 `event_date` 值，同时 `end_time_obj` 将保存当前记录的 `event_date` 值。

当给定的地点经过第一个迭代之后，`start_time_obj` 将不再是 `None`，此时可以对两个 `datetime` 对象进行 diff 比较。`end_time_obj` 减去 `start_time_obj` 得到一个时间差值对象，如果这个值比 `current_diff` 大，将 `current_diff` 设置为这个值。这样做之后，得到了对于这个地点的发生事件之间最长间隔时间区间。一旦处理完所有关于这个地点的数据，设置 `start_date` 以及 `end_date` 用于以后的输出。和之前提及的一样，无论 `current_diff` 是否发生变化，都需要将 `prev_date` 设置为 `event_date`，将 `start_time_obj` 设置为 `end_time_obj`。

当下一次进入循环时，如果 `loc` 不等于 `current_loc`，则在进入下一步之前，输出当前的最长的事件时间区间。从 `stdout` 输出的每行记录被写入 Hive 表。其中 `current_loc`

表示当前地点，start_date 表示 event_date 的下界，end_date 表示 event_date 的上界，current_diff.days 表示这两个日期之间的间隔天数。

更多参考

下面列出了本节涉及的一些操作的更多内容。

SORT BY、DISTRIBUTE BY、CLUSTER BY 以及 ORDER BY

这 4 种操作时常会引起 Hive 初学者的疑惑。下面对这四种操作进行快速比较，这样你可以根据你的情况选择正确的操作。

- DISTRIBUT BY：根据列的值，对应的行将被分配至同一个 reducer 处。如果单独使用，将不保证输入至 reducer 的顺序。
- SORT BY：表明输入至 reducer 的记录，根据指定的列进行排序
- CLUSTER BY：这是对于一组列集合同时执行 SORT BY 以及 DISTRIBUTE BY 的操作的简写。
- ORDER BY：与传统的 SQL 操作类似。该操作保持每个 reducer 输出的记录有序。小心使用该操作，因为这将强制只使用一个 reducer 进行排序并输出记录。强烈推荐使用 LIMIT 语句。

关键字 MAP、REDUCE 是 SELECT TRANSFORM 的简写

Hive 关键字 MAP 以及 REDUCE 是 SELECT TRANSFORM 的简写。并非强制查询的执行过程在各个阶段切换。你可以使用三个中的任意一个达到同样的结果。这些设定都是为了使查询具有更好的可读性。

延伸阅读

- 第 3 章的"使用 Hive 及 Python 清洗、转换地理事件数据"（3.8 节）。

6.7 使用 Pig 计算 Audioscrobbler 数据集中艺术家之间的余弦相似度

余弦相似度用于度量两个向量的相似程度。本节根据 Audioscrobbler 的用户添加艺术家至播放列表的次数，使用余弦相似度寻找相似的艺术家。方法是计算用户既播放 artist 1

又播放 artist 2 的频率。

准备工作

从 http://www.packtpub.com/support 下载 Audioscrobbler 数据集。

操作步骤

执行下面的步骤，使用 Pig 计算余弦相似度。

1. 复制文件 artist_data.txt、user_artist_data.txt 至 HDFS：

```
hadoop fs -put artist_data.txt user_artist_data.txt /data/audioscrobbler/
```

2. 加载数据至 Pig：

```
plays = load '/data/audioscrobbler/user_artist_data.txt'
        using PigStorage(' ') as (user_id:long, artist_id:long,
playcount:long);

artist = load '/data/audioscrobbler/artist_data.txt' as (artist_
id:long, artist_name:chararray);
```

3. 采样文件 user_artist_data.txt：

```
plays = sample plays .01;
```

4. 规约化播放次数从 0 至 100：

```
user_total_grp = group plays by user_id;

user_total = foreach user_total_grp generate group as user_id,
SUM(plays.playcount) as totalplays;

plays_user_total = join plays by user_id, user_total by user_id
using 'replicated';

norm_plays = foreach plays_user_total generate user_total::user_id
as user_id, artist_id, ((double)playcount/(double)totalplays) *
100.0 as norm_play_cnt;
```

5. 获得每个用户的艺术家对：

```
norm_plays2 = foreach norm_plays generate *;

play_pairs = join norm_plays by user_id, norm_plays2 by user_id
using 'replicated';

play_pairs = filter play_pairs by norm_plays::plays::artist_id !=
norm_plays2::plays::artist_id;
```

6. 计算余弦相似度：

```
cos_sim_step1 = foreach play_pairs generate ((double)norm_
plays::norm_play_cnt) * (double)norm_plays2::norm_play_cnt) as
dot_product_step1, ((double)norm_plays::norm_play_cnt *(double)
norm_plays::norm_play_cnt) as play1_sq;
((double)norm_plays2::norm_play_cnt *(double) norm_plays2::norm_
play_cnt) as play2_sq;

cos_sim_grp = group cos_sim_step1 by (norm_plays::plays::artist_
id, norm_plays2::plays::artist_id);

cos_sim_step2 = foreach cos_sim_grp generate flatten(group),
COUNT(cos_sim_step1.dot_prodct_step1) as cnt, SUM(cos_sim_step1.
dot_product_step1) as dot_product, SUM(cos_sim_step1.norm_
plays::norm_play_cnt) as tot_play_sq, SUM(cos_sim_step1.norm_
plays2::norm_play_cnt) as tot_play_sq2;

cos_sim = foreach cos_sim_step2 generate group::norm_
plays::plays::artist_id as artist_id1, group::norm_plays2::plays_
artist_id as artist_id2, dot_product /sqr (tot_play_sq1 * tot_play_
sq2) as cosine_similarity;①
```

7. 获得艺术家的名字：

```
art1 = join cos_sim by artist_id1, artist by artist_id using 'replicated';
art2 = join art1 by artist_id2, artist by artist_id using 'replicated';
art3 = foreach art2 generate artist_id1, art1::artist::artist_name
as artist_name1, artist_id2, artist::artist_name as artist_name2, cosin_
similarity;
```

8. 输出相似度最高的 25 条记录：

```
top = order art3 by cosine_similarity DESC;
top_25 = limit top 25;
dump top25;
```

输出如下：

```
(1000157,AC/DC,3418,Hole,0.9115799166673817)
(829,Nas,1002216,The Darkness,0.9110152004952198)
(1022845,Jessica Simpson,1002325,Mandy Moore,0.9097097460071537)
(53,Wu-Tang Clan,78,Sublime,0.9096468367168238)
(1001180,Godsmack,1234871,Devildriver,0.9093019011575069)
```

---

① 这里的余弦相似度实现有问题，应该是分母还要开根号！

```
(1001594,Adema,1007903,Maroon 5,0.909297052154195)
(689,Bette Midler,1003904,Better Than Ezra,0.9089467492461345)
(949,Ben Folds Five,2745,Ladytron,0.908736095810886)
(1000388,Ben Folds,930,Eminem,0.9085664586931873)
(1013654,Who Da Funk,5672,Nancy Sinatra,0.9084521262343653)
(1005386,Stabbing Westward,30,Jane's Addiction,0.9075360259222892)
(1252,Travis,1275996,R.E.M.,0.9071980963712077)
(100,Phoenix,1278,Ryan Adams,0.9071754511713067)
(2247,Four Tet,1009898,A Silver Mt. Zion,0.9069623744896833)
(1037970,Kanye West,1000991,Alison Krauss,0.9058717234023009)
(352,Beck,5672,Nancy Sinatra,0.9056851798338253)
(831,Nine Inch Nails,1251,Morcheeba,0.9051453756031981)
(1007004,Journey,1005479,Mr. Mister,0.9041311825160151)
(1002470,Elton John,1000416,Ramones,0.9040551837635081)
(1200,Faith No More,1007903,Maroon 5,0.9038274644717641)
(1002850,Glassjaw,1016435,Senses Fail,0.9034604126636377)
(1004294,Thursday,2439,HiM,0.902728300518356)
(1003259,ABBA,1057704,Readymade,0.9026955950032872)
(1001590,Hybrid,791,Beenie Man,0.9020872203833108)
(1501,Wolfgang Amadeus Mozart,4569,Simon &
Garfunkel,0.9018860912385024)
```

工作原理

`load` 语句定义导入 Pig 的数据的格式与类型。Pig 是惰性加载数据，这意味着一开始加载语句不会执行任何操作，直到加载需要输出数据的语句。

对文件 `user_artist_data.txt` 进行采样，这样可以使用复制连接对其进行自连接。这样做显然以准确性为代价，降低了处理时间。采样率这里设置为 0.01，表明大概有百分之一的数据行被加载。

将用户选择播放一个艺术家的作品视为对艺术家的一次投票。播放次数被规范化至 100，从而保证了每个用户能投票的次数都一样。

根据 `user_id` 对文件 `user_artist_data.txt` 进行自连接，生成了用户添加至播放列表的所有的艺术家对。过滤器去掉了由于自连接导致重复的记录。

接下来的语句用于计算余弦相似度。对于用户添加至播放列表的每个艺术家对，将 artist 1 的播放次数与 artist 2 的播放次数相乘。接下来输出 artist 1 的播放次数与 artist 2 的播放次数。按照每个艺术家对聚集之前的结果。对每个用户的 artist1 播放数与 artist 2 播放数的乘积求和，作为两者的点积。分别对所有用户播放的 artist1 和 artist2 求和。余弦相似度为计算的点积除以 artist1 总的播放量与 artist2 总的播放量的乘积。方法是计算出同时播放 artist1 和 artist2 的用户的频率。

6.8 使用 Pig 以及 datafu 剔除 Audioscrobbler 数据集中的离群值

Datafu 是 LinkedIn 公司 SNA 小组开发的 Pig UDF 开源函数库，其中包含大量有用的函数。本节使用 Audioscrobbler 数据集的**播放次数**以及 datafu 提供的 **Quantile** UDF，识别并剔除离群值。

准备工作

- 从 `https://github.com/linkedin/datafu/downloads` 下载 datfu 版本 0.0.4。
- 解压缩 TAR 文件。添加文件 `datafu-0.0.4/dist/datafu-0.0.4.jar` 至 Pig 能访问的路径。
- 从 `http://www.packtpub.com/support` 下载 Audioscrobbler 数据集。

操作步骤

1. 注册 datafu JAR 文件，创建 `Quantile` UDF：

```
register /path/to/datafu-0.0.4.jar;
define Quantile datafu.pig.stats.Quantile('.90');
```

2. 读取文件 user_artist_data.txt：

```
plays = load '/data/audioscrobbler.txt'using PigStorage(' ') as
(user_id:long, artist_id:long, playcount:long);
```

3. 聚集所有的数据：

```
plays_grp = group plays ALL;
```

4. 生成 90%，作为判断离群值的最大值：

```
out_max = foreach plays_grp{
        ord = order plays by playcount;
        generate Quantile(ord.playcount) as ninetieth ;
        }
```

5. 过滤大于 90%的离群值：

```
trim_outliers = foreach plays generate user_id, artist_id,
(playcount>out_max.ninetieth ? out_max.ninetieth : playcount);
```

6. 存储过滤离群值的 `user_artist_data.txt` 文件：

```
store trim_outliers into '/data/audioscrobble/outliers_trimmed.bcp';
```

工作原理

本节利用 LinkedIn 提供的 datafu 开源库。一旦注册了 JAR 文件，可以在 Pig 脚本中使用 JAR 创建的 UDF。define 命令将 .90 作为参数，调用 datafu.pig.stats.Quantile UDF 的构造函数。Quantile UDF 的构造函数将会创建一个实例，用于生成输入向量的百分之九十分位数。同时 define 定义 Quantile 为该 UDF 的一个简写。

user-artist 数据集加载至命名为 plays 的 Pig 关系。该数据通过 ALL 进行聚集。组 ALL 是一个特殊的组，将所有的输入包含至一个 bag 中。

Quantile UDF 需要输入的数据首先是有序的。这里的数据是按照播放次数排序，已排好序的播放次数向量作为 Quantile UDF 的输入。有序的播放次数数据简化了 Quantile UDF 的工作。此时将该值的百分之九十分位数输出。

接下来，该值与 user-artist 文件中的每个播放次数比较，如果这个播放次数比 Quantile UDF 返回的值更大，将剔除这个值，否者将保留这个值。

剔除掉离群值的 user-artist 文件将保存至 HDFS，用来进行后续的处理。

更多参考

Datafu 库同样包含 StreamingQuanitle UDF。这个 UDF 与 Quantile UDF 类似，不同的是这个 UDF 不需要使用的数据已经排序。这将大大提高操作的性能，可是也会带来一些代价，那就是 StreamingQuantile UDF 只提供一个估计值。

```
define Quantile datafu.pig.stats.StreamingQuantile('.90');
```

第 7 章

高级大数据分析

本章我们将介绍：
- 使用 Apache Giraph 计算 PageRank
- 使用 Apache Giraph 计算单源最短路径
- 使用 Apache Giraph 执行分布式宽度优先搜索
- 使用 Apache Mahout 计算协同过滤
- 使用 Apache Mahout 进行聚类
- 使用 Apache Mahout 进行情感分类

7.1 介绍

使用 MapReduce 框架很难解决图计算和机器学习的问题。大多数这些问题需要使用复杂的算法进行多步迭代，通过 MapReduce 实现是十分麻烦的。幸运的是，有两个有用的框架可以在 Hadoop 环境中帮助解决图计算和机器学习的问题。Apache **Giraph** 是一个为了运行大规模算法而设计的图处理框架。Apache **Mahout** 框架提供了关于分布式机器学习算法的实现。

本章将向读者介绍这两个框架，这些框架都充分利用了 MapReduce 分布式处理的能力。

7.2 使用 Apache Giraph 计算 PageRank

本节的首要目的在于建立和测试 Apache Giraph 默认提供的基于 Google Pregel 建模而实现的

PageRank 的例子。这里将展示在一个伪分布式 Hadoop 集群，提交和执行 Giraph 作业的步骤。

准备工作

对于第一次运行 Giraph 的用户，我们建议在伪分布式 Hadoop 集群上运行本节的内容。

对于客户机而言，需要在用户环境路径中安装并配置 SVN 以及 Maven。

本节不需要对 Giraph API 有一个全面的了解，但是需要对**大容量同步并行处理（BSP）**以及以节点为中心的 API 设计（例如 Apache Giraph 以及 Google Pregel）有一定了解。

操作步骤

执行下面的步骤，建立并测试默认的 Giraph PageRank 的例子。

1. 进入一个默认的文件夹，从 Apache 官方网站 checkout 最新版的 Giraph 源代码：

```
$ svn co https://svn.apache.org/repos/asf/giraph/trunk
```

2. 在这个文件夹，编译 trunk 代码：

```
$ mvn compile
```

3. 一旦编译完成，在 trunk 对应的目标文件夹中可以找到 JAR 文件 `giraph-0.2-SNAPSHOT-jar-with-dependencies.jar`。

4. 执行如下命令：

```
hadoop jar giraph-0.2-SNAPSHOT-jar-with-dependencies.jar org.apache.giraph.benchmark.PageRankBenchmark -V 1000 -e 1 -s 5 -w 1 -V
```

5. 作业将会开始执行，MapReduce 命令行会输出成功的提示。

6. 输出的 Giraph 状态的计数器，如下所示：

```
INFO mapred.JobClient:    Giraph Stats
INFO mapred.JobClient:        Aggregate edges=1000
mapred.JobClient:        Superstep=6
mapred.JobClient:        Last checkpointed superstep=0
mapred.JobClient:        Current workers=1
mapred.JobClient:        Current master task partition=0
mapred.JobClient:        Sent messages=0
mapred.JobClient:        Aggregate finished vertices=1000
mapred.JobClient:        Aggregate vertices=1000
```

工作原理

首先，使用 SVN 从 Apache 官方网站 checkout 最新版的源代码。一旦编译 JAR 文件完成，这个 `PageRankBenchmark` 的例子是可以直接提交的。在开始测试 Giraph 之前，需要设置如下的命令行参数。

- ❑ -v：运行 PageRank 的节点总数。选择 1000 作为测试。对于更加准确地测试可以在完全分布式环境上测试上百万个节点。

- ❑ -e：对于每个节点，出度边的个数。这将控制在每个大步（super step）中输出至相邻节点的消息数目，这里邻居节点定义为通过一条或多条边连接另一个节点的节点。

- ❑ -s：在结束 PageRank 计算之前，大步的运行次数。

- ❑ -w：处理每个图划分的 worker 的总数。由于我们运行在伪分布式集群中（单一主机），将其限制为 1 是安全的做法。在一个全分布式集群中，需要在不同的物理主机分布多个 worker。

- ❑ -v：以详尽模式，在命令行输出作业的执行过程。

这个作业除了核心的 Hadoop/ZooKeeper 依赖之外，没有其他的 classpath 依赖。通过在命令行直接执行 `hadoop jar` 就能将作业直接提交至集群。

例子 `PageRankBenchmark` 不将任何结果写回至 HDFS。这样首要的作用是测试并揭示集群执行 Giraph 作业可能存在的瓶颈。对于执行包含大量节点并包含多条边的作业，将会出现内存瓶颈、worker 之间的网络 I/O 连接以及其他隐藏的问题。

更多参考

Apache Giraph 是一个相对较新的开源迭代图处理计算框架。下面的提示将更有助于加深对其的理解。

紧跟 Apache Giraph 社区步伐

Apache Giraph 有一个特别活跃的开发社区。API 时常增加新的功能，进行 bug 修复，有时也会重构。至少每周从主干更新你的代码是一个好的建议。在书写本书的时候，Giraph 还没有公开的 Maven 工程。这将在近期改变，对于目前而言需要通过 SVN 进行代码的更新。

阅读、理解关于 Google Pregel 的论文

2009 年，Google 发表了一篇描述他们软件系统技术细节的高水平研究论文，主要是基于大容量同步并行处理（BSP）模型构建了大规模处理以图为中心的系统。

Apache Giraph 是对这篇论文众多概念的一个开源实现。熟悉 Pregel 的设计有助于理解 Giraph 代码中的众多组成部分。

了解 BSP 的基本概念请访问：`http://en.wikipedia.org/wiki/Bulk_Synchronous_Parallel`。

延伸阅读

- 使用 Apache Giraph 计算单源最短路径（7.3 节）。
- 使用 Apache Giraph 执行分布式宽度优先搜索（7.4 节）。

7.3 使用 Apache Giraph 计算单源最短路径

本节我们将在非循环有向图连接的雇员关系中，实现 Google Pregel 最短路径的一个变形。代码将对图中的每个节点赋予一个唯一的 ID 值，并标记从一个源 ID 节点到达其他节点需要的最小跳转次数。在 HDFS 存储的雇员网络关系数据表示为以横线分割的 RDF 三元组。

资源描述框架（RDF）是一种表示实体以及实体之间关系的有效的数据格式。

准备工作

确保你已经对 Google Pregel/BSP 以及 Giraph API 有一定的了解。

现在需要访问一个伪分布式 Hadoop 集群。本节的代码使用非 split 的 master worker 配置项。这个配置项对于全分布式环境是不适合的。这里同时假设你对 bash shell 脚本也同样熟悉。

你需要将示例数据集 gooftech.tsv 放置于 HDFS 文件夹 /input/gooftech。

同时，你需要将这些代码打包至一个 JAR 文件中，并通过 shell 运行 Hadoop JAR launcher 执行该代码。本节列出的 shell 脚本表示了一个包含正确的 classpath 依赖的作业提交的模板。

操作步骤

执行下面的步骤，使用 Giraph 实现最短路径算法。

1. 首先，通过继承类 TextInputFormat 定义自定义类 InputFormat，从本文中读取雇员 RDF 三元组。在你所选定的包保存 EmployeeRDFTextInputFormat.java 类：

```
import com.google.common.collect.Maps;
import org.apache.giraph.graph.BspUtils;
import org.apache.giraph.graph.Vertex;
import org.apache.giraph.graph.VertexReader;
import org.apache.giraph.lib.TextVertexInputFormat;
import org.apache.hadoop.io.*;
import org.apache.hadoop.mapreduce.InputSplit;
import org.apache.hadoop.mapreduce.RecordReader;
import org.apache.hadoop.mapreduce.TaskAttemptContext;

import java.io.IOException;
```

```java
import java.util.Map;
import java.util.regex.Pattern;

public class EmployeeRDFTextInputFormat extends TextVertexInputFormat<Text,
IntWritable, NullWritable, IntWritable> {

  @Override
  public VertexReader<Text, IntWritable, NullWritable, IntWritable>
  createVertexReader(InputSplit split, TaskAttemptContext context)
    throws IOException {
    return new EmployeeRDFVertexReader(
        textInputFormat.createRecordReader(split, context));
  }
}
```

2. 创建自定义节点 reader 静态内部类,处理输入格式:

```java
    public static class EmployeeRDFVertexReader extends
        TextVertexInputFormat.TextVertexReader<Text, IntWritable, NullWritable,
IntWritable> {

        private static final Pattern TAB = Pattern.compile("[\\t]");
        private static final Pattern COLON = Pattern.compile("[:]");
        private static final Pattern COMMA = Pattern.compile("[,]");

        public EmployeeRDFVertexReader(RecordReader<LongWritable, Text>
lineReader) {
            super(lineReader);
        }
```

3. 重写方法 getCurrentVertex()。这个方法是行 reader 用来解析自定义节点对象的:

```java
        @Override
        public Vertex<Text, IntWritable, NullWritable, IntWritable>
        getCurrentVertex() throws IOException, InterruptedException {
          Vertex<Text, IntWritable, NullWritable, IntWritable>
          vertex = BspUtils.<Text, IntWritable, NullWritable, IntWritable>
          createVertex(getContext().getConfiguration());

          String[] tokens = TAB.split(getRecordReader()
              .getCurrentValue().toString());
          Text vertexId = new Text(tokens[0]);

          IntWritable value = new IntWritable(0);
          String subtoken = COLON.split(tokens[2])[1];
          String[] subs = COMMA.split(subtoken);
          Map<Text, NullWritable> edges =
              Maps.newHashMapWithExpectedSize(subs.length);
          for(String sub : subs) {
             if(!sub.equals("none"))
                 edges.put(new Text(sub), NullWritable.get());
          }

          vertex.initialize(vertexId, value, edges, null);
            return vertex;
```

```
            }

            @Override
            public boolean nextVertex() throws IOException, InterruptedException {
                return getRecordReader().nextKeyValue();
            }
        }
    }
```

4. 作业的启动部分代码、节点类以及自定义输出格式都包含在一个类中。在你选择的包中，保存如下的代码，并命名为 EmployeeShortestPath.java：

```
import org.apache.giraph.graph.*;
import org.apache.giraph.lib.TextVertexOutputFormat;
import org.apache.hadoop.conf.Configuration;
import org.apache.hadoop.fs.FileSystem;
import org.apache.hadoop.fs.Path;
import org.apache.hadoop.io.*;
import org.apache.hadoop.mapreduce.RecordWriter;
import org.apache.hadoop.mapreduce.TaskAttemptContext;
import org.apache.hadoop.mapreduce.lib.input.FileInputFormat;
import org.apache.hadoop.mapreduce.lib.output.FileOutputFormat;
import org.apache.hadoop.util.Tool;
import org.apache.hadoop.util.ToolRunner;

import java.io.IOException;

/**
 * Value based on number of hops. vertices receiving incoming
messages increment the message
 */
public class EmployeeShortestPath implements Tool{

    public static final String NAME = "emp_shortest_path";

    private Configuration conf;
    private static final String SOURCE_ID = "emp_source_id";

    public EmployeeShortestPath(Configuration configuration) {
        conf = configuration;
    }
```

5. 下面的 run() 方法设置 Giraph 作业的配置项：

```
    @Override
    public int run(String[] args) throws Exception {
        if(args.length < 4) {
            System.err.println(printUsage());
            System.exit(1);
        }
        if(args.length > 4) {
            System.err.println("too many arguments. " + "Did you forget to quote the source ID name ('firstname lastname')");
```

```
            System.exit(1);
        }
        String input = args[0];
        String output = args[1];
        String source_id = args[2];
        String zooQuorum = args[3];

        conf.set(SOURCE_ID, source_id);
        conf.setBoolean(GiraphJob.SPLIT_MASTER_WORKER, false);
        conf.setBoolean(GiraphJob.USE_SUPERSTEP_COUNTERS, false);
        conf.setInt(GiraphJob.CHECKPOINT_FREQUENCY, 0);
        GiraphJob job = new GiraphJob(conf, "single-source
          shortest path for employee: " + source_id);
        job.setVertexClass(EmployeeShortestPathVertex.class);
        job.setVertexInputFormatClass(EmployeeRDFTextInputFormat.class);
     job.setVertexOutputFormatClass(EmployeeShortestPathOutputFormat.class);
        job.setZooKeeperConfiguration(zooQuorum);

        FileInputFormat.addInputPath(job.getInternalJob(), new Path(input));
        FileOutputFormat.setOutputPath(job.getInternalJob(), removeAndSet
Output(output));

        job.setWorkerConfiguration(1, 1, 100.0f);
        return job.run(true) ? 0 : 1;
    }
```

6. 如下的代码，强行删除已存在 HDFS 的输出目录。请小心使用。其他的方法需要和 Tool 接口一致：

```
    private Path removeAndSetOutput(String outputDir) throws IOException {
        FileSystem fs = FileSystem.get(conf);
        Path path = new Path(outputDir);
        fs.delete(path, true);
        return path;
    }


    private String printUsage() {
        return "usage: <input> <output> <single quoted source_id> <zookeeper_
quorum>";
    }

    @Override
    public void setConf(Configuration conf) {
        this.conf = conf;
    }

    @Override
    public Configuration getConf() {
        return conf;
    }
```

7. main()方法使用 ToolRunner 进行实例化并提交该作业:

```java
public static void main(String[] args) throws Exception {
    System.exit(ToolRunner.run(new EmployeeShortestPath(new Configuration()),
args));
}
```

8. 静态内部类 EmployeeShortestPathVertex 为每个大步过程定义了一个用户自定义的计算方法:

```java
public static class EmployeeShortestPathVertex<I extends WritableComparable,
V extends Writable, E extends Writable, M extends Writable> extends EdgeListVertex
<Text, IntWritable, NullWritable,
IntWritable>
{

    private IntWritable max = new IntWritable(Integer.MAX_VALUE);
    private IntWritable msg = new IntWritable(1);

    private boolean isSource() {
        return getId().toString().equals(
            getConf().get(SOURCE_ID));
    }

    @Override
    public void compute(Iterable<IntWritable> messages)
                    throws IOException {
        if(getSuperstep() == 0) {
            setValue(max);
            if(isSource()) {
                for(Edge<Text, NullWritable> e :
                    getEdges()) {
                    sendMessage(e.getTargetVertexId(),
                        msg);
                }
            }
        }
        int min = getValue().get();
        for(IntWritable msg : messages) {
            min = Math.min(msg.get(), min);
        }
        if(min < getValue().get()) {
            setValue(new IntWritable(min));
            msg.set(min + 1);
            sendMessageToAllEdges(msg);
        }
        voteToHalt();
    }
}
```

9. 内部静态类 EmployeeShortestPathOutputFormat 定义了用户自定义 OutputFormat。类 EmployeeRDFVertexWriter 按照 Text 的键值对输出节点信息到 HDFS 中。

```java
    public static class EmployeeShortestPathOutputFormat extends
TextVertexOutputFormat <Text, IntWritable, NullWritable> {

        private static class EmployeeRDFVertexWriter
                extends TextVertexWriter <Text, IntWritable, NullWritable> {
          private Text valOut = new Text();

          public EmployeeRDFVertexWriter(
                  RecordWriter<Text, Text> lineRecordWriter) {
              super(lineRecordWriter);
          }

          @Override
          public void writeVertex(
                  Vertex<Text, IntWritable, NullWritable,?> vertex)
                  throws IOException,
                              InterruptedException {

              valOut.set(vertex.getValue().toString());
              if(vertex.getValue().get() == Integer.MAX_VALUE)
                  valOut.set("no path");
              getRecordWriter().write(vertex.getId(), valOut);
          }

        }
        @Override
        public VertexWriter<Text, IntWritable, NullWritable>
        createVertexWriter(TaskAttemptContext context)
              throws IOException, InterruptedException {
          RecordWriter<Text, Text> recordWriter =
                  textOutputFormat.getRecordWriter(context);
          return new EmployeeRDFVertexWriter(recordWriter);
        }
    }
}
```

10. 创建 shell 脚本 run_employee_shortest_path.sh,其中的代码命令行如下所示。将 GIRAPH_PATH 修改为 Giraph JAR 文件的本地路径,将 JAR_PATH 修改为刚刚编译的自定义 JAR 文件的本地路径。

 为了在自定义 JAR 文件中使用别名 emp_shortest_path,必须在主类中使用 Hadoop 提供的 Driver 类。

```
GIRAPH_PATH=lib/giraph/giraph-0.2-SNAPSHOT-jar-with-dependencies.jar
HADOOP_CLASSPATH=$HADOOP_CLASSPATH:$GIRAPH_PATH
JAR_PATH=dist/employee_examples.jar
export HADOOP_CLASSPATH
  hadoop jar $JAR_PATH emp_shortest_path -libjars $GIRAPH_PATH,$JAR_PATH /input/gooftech /output/gooftech 'Shanae Dailey' localhost:2181
```

11. 执行 run_employee_shortest_path.sh。这个作业将会被提交至 Hadoop 集群。

在 /output/gooftech 会得到一个单 part 的文件，其中包含了从源 ID 到每个雇员需要的最短路径长度，如果一个雇员是不可达的则会显示 no path。

工作原理

我们首先分析用户自定义输入格式。Giraph API 提供的 TextVertexInputFormat 封装了用于读取存储在文本文件中每行节点信息的类 TextInputFormat 以及类 LineReader。目前，Giraph API 需要记录按照节点 ID 进行排序。这里的雇员数据集是按照 firstname/lastname 排序的，因此可以满足 Giraph 的要求，能直接进行下一步。为了从 RDF 数据中生成有意义的节点，创建了 TextVertexInputFormat 的子类 EmployeeRDFTextInputFormat。同时，为了准确地控制节点的出现形式，创建了 TextVertexReader 的子类 EmployeeRDFVertexReader。这需要在自定义输入格式类重载 getRecordReader() 方法，返回自定义 reader 子类的一个实例。记录 reader 代表一个 Hadoop LineReader 的实例，从每个输入分片文件的文本行中创建节点对象。这里，重载了方法 getCurrentVertex()，从行 reader 得到的 RDF 元组生成不同的节点对象。由于继承了 TextVertexReader，不需要担心对每一行记录手动地控制调用 getCurrentVertex() 方法，框架会自动完成相关工作，只需要告诉框架如何将每行文本记录转化为包含一条或多条边的节点。

代码中，声明了函数 EmployeeRDFTextInputFormat 的泛型参数，从左到右分别提供了具体节点 ID 信息类，节点类，边类以及信息类。下面函数的泛型声明在父类中被定义：

```
public abstract class TextVertexInputFormat<I extends WritableComparable,V extends Writable, E extends Writable, M extends Writable> extends VertexInputFormat<I, V, E, M>
```

四个泛型类型必须为 Writable。节点 ID 类则必须是 WritableComparable。目前，Giraph 框架还不支持其他的序列化框架。

这里对 getCurrentVertex() 方法进行了简单的实现。我们定义了一些用于正确分割 RDF 元组的静态常量正则模式。合并 firstname/lastname 作为节点的 ID 并存储为一个 Text 实例。每个节点的值字段的值初始化为 IntWriable 类型的 0。由逗号分割的每个列表组成元素都是一个边 ID 的引用。可是由于我们不需要每条边的值信息，将边的值定义为 NullWritable 就可以满足需求。对于这个特定的作业，消息类型是 IntWritable。这个类型将在 7.4 节中得到复用。为了精简篇幅内容，只在这里对输入类型进行详细解释。

接下来，配置作业类。作业的启动严重依赖 Hadoop MapReduce 的 API。我们实现了 Tool 接口，从命令行读取数据定义了四个参数值。分别是，作业需要的 HDFS 输入文件目录，写回至 HDFS 的输出文件目录，执行单源最短路径的源 ID 号，以及管理作业状态的 ZooKeeper quorum 参数。随后，由于运行在资源受限的伪分布式环境下，需要定义一些其他的参数。

```
conf.setBoolean(GiraphJob.SPLIT_MASTER_WORKER, false);
conf.setBoolean(GiraphJob.USE_SUPERSTEP_COUNTERS, false);
conf.setInt(GiraphJob.CHECKPOINT_FREQUENCY, 0);
```

`SPLIT_MASTER_WORKER` 告知 Giraph 的 master 是否在不同的主机上运行 worker。默认的这个值设置为 `true`，但是由于这里运行在伪分布式环境的节点上，需要将其设置为 `false`。关闭大步计数器将减少作业 MapReduce 的 WebUI 相关冗余项的显示。对于测试包含成百上千大步的作业是很方便的。最后，关闭检测点，这样 Giraph 将会知道不需要关心备份任何大步的图状态。这样做是因为这只是在测试，关注的是快速地执行作业。作为产品级的作业，虽然会降低整个作业的运行时间，但通常建议开启图状态检测点。之后，初始化了一个 GiraphJob 实例，需要设置一些关于这个作业实例描述性标题的配置项。

接下来的三行对于在集群上正确的执行 Giraph 作业是至关重要的：

```
job.setVertexClass(EmployeeShortestPathVertex.class);
job.setVertexInputFormatClass(EmployeeRDFTextInputFormat.class);
job.setVertexOutputFormatClass(EmployeeShortestPathOutputFormat.class);
```

第一行告诉 Giraph 实现了封装图中每个节点的自定义节点类。这里面包括每个大步调用的与应用相关的 `compute()` 方法。我们继承了基类 `EdgeListVertex`，利用之前存在的一些代码实现消息处理、边迭代以及成员变量序列化。

接着，设置 ZooKeeper quorum，定义了单个 worker 处理图的一个划分。如果你的伪分布式集群能支持多个 worker(多并发 map JVM)，可以增加这个上限值。只需要注意为 master 进程保留一个空闲的 map 槽位。最后，已经准备好将这个作业提交至集群。

在 `InputFormat` 从不同分片文件中创建完节点对象之后，每个节点的 `compute()` 方法将会被调用。定义的静态内部类 `EmployeeShortestPathVertex` 重载了 `compute()` 方法，实现了计算最短路径的业务逻辑。特别的，我们对图中从源节点到每个其他连接的节点的最小跳转数更感兴趣，如果目标节点到源节点是非连通的，则返回 `no path`。

第一大步（S0）

在 S0 阶段，函数立即通过第一个判断条件，初始化每个节点值为最大的整数值。当收到传递的消息时，每个节点比较当前值与收到的消息的值，并保留更低的值。因此对于该业务逻辑，将最小的值初始化为该数据类型的最大值会简化该逻辑。在 S0 过程中，重要的是源节点发送一条消息至与之相连的其他节点，这些节点到源节点需要一次跳转。为了完成这项操作，定义了一个成员实例 `msg`，用来表示消息对象。每次节点需要发送一个消息的时候，这个对象将会被重置与复用，这样避免了不必要的实例化过程。

需要比较当前收到的任何消息以及目前保存的最小跳转数，如果需要更新这个值，需要通知相关的边。由于只位于 S0 阶段，没有任何消息，因此这个值为 Integer.MAX。因为这个最小值不会变化，所以避免了进入最后的条件分支。

在作业的每个大步的最后都将调用方法 voteToHalt()。Giraph 框架在下一个大步阶段收到消息，将自动激活该节点，而对于暂时完成发送/接收消息的节点，我们需要将其置为非激活状态。一旦图中的任何节点不再处理任何消息时，作业将停止重新激活节点的操作，并且作业将认为已经完成。

第二大步（S1）

在上一个大步执行完之后，每个图中的单节点会投票挂起作业的执行。唯一发送消息至它相关边的节点是源节点。因此，框架将会重新激活与源节点相连接的节点。源节点告诉连接边之间只有一跳，这个值比 Integer.MAX 小，将会立刻取代节点当前的保存值。每个收到消息的节点立即通知与它连接的边，离源节点为 min+1 跳。上面的过程循环执行下去。

当任何连接的边收到的消息低于当前的节点值，表明从源 ID 到当前节点有一条更低跳转数的路径。此时需要重新通知连接当前节点的边。

最后，每个节点得到它离源 ID 的最短距离，在当前大步的第 N 步，不再有任何消息发送。当开始大步的第 N+1 步，没有需要用来处理收到消息的激活节点，从而完成整个作业。此时，需要输出每个节点到源节点的最小跳转数。

为了将节点值信息作为文本写回至 HDFS，我们实现了类 TextVertexOutputFormat 的静态内部子类 EmployeeShortestPathOutputFormat。这个类与我们之前自定义的 InputFormat 的继承/委托模式类似，不同的是委托模式使用自定义类 RecordWriter 代替自定义类 RecordReader。我们设置了一个 Text 类型的成员变量 valOut，将整型数字作为一个字符串进行重用。框架自动处理包含在数据集中每个节点的 writeVertex() 方法的调用。

如果当前节点值仍然等于 Integer.MAX，此时图没有收到任何来自这个节点的消息，表明图对于源节点是不可达的。否则，输出源节点 ID 到达当前节点 ID 的最小跳转数。

延伸阅读

- 使用 Apache Giraph 执行分布式宽度优先搜索（7.4 节）。

7.4 使用 Apache Giraph 执行分布式宽度优先搜索

本节中使用 Apache Giraph API 实现分布式深度优先搜索，确定两个雇员通过公司网络的一条或多条路径是否连接。代码的功能依赖于雇员节点之间的消息，决定了节点是否可达。

准备工作

确保你已经对 Google Pregel/BSP 以及 Giraph API 有一定的了解。

现在需要访问一个伪分布式 Hadoop 集群。本节的代码使用不包含 split 的 master worker 配置项。这个配置项对于全分布式环境是不适合的。这里同时假设你对于 bash shell 脚本也同样熟悉。

需要将示例数据集 gooftech.tsv 放置于 HDFS 文件夹 /input/gooftech。

同时,你需要将这些代码打包至一个 JAR 文件中,并通过 shell 运行 Hadoop JAR launcher 执行该代码。本节列出的 shell 脚本表示了一个包含正确的 classpath 依赖的作业提交的模板。

操作步骤

执行下面的步骤,使用 Giraph 实现宽度优先搜索。

1. 实现 `EmployeeRDFTextInputFormat.java`。具体实现参见 7.3 节的"操作步骤"部分。

2. 作业的启动部分代码、节点类以及自定义输出格式都包含在一个类中。在你选择的包中,保存如下的代码,并命名为 `EmployeeBreadthFirstSearch.java`:

```java
import org.apache.giraph.graph.*;
import org.apache.giraph.lib.TextVertexOutputFormat;
import org.apache.hadoop.conf.Configuration;
import org.apache.hadoop.fs.FileSystem;
import org.apache.hadoop.fs.Path;
import org.apache.hadoop.io.*;
import org.apache.hadoop.mapreduce.RecordWriter;
import org.apache.hadoop.mapreduce.TaskAttemptContext;
import org.apache.hadoop.mapreduce.lib.input.FileInputFormat;
import org.apache.hadoop.mapreduce.lib.output.FileOutputFormat;
import org.apache.hadoop.util.Tool;
import org.apache.hadoop.util.ToolRunner;

import java.io.IOException;

/**
 * Start with specified employee, mark the target if message is
 received
 */
public class EmployeeBreadthFirstSearch implements Tool{

    public static final String NAME = "emp_breadth_search";
```

```java
    private Configuration conf;
    private static final String SOURCE_ID = "emp_src_id";
    private static final String DEST_ID = "emp_dest_id";

    public EmployeeBreadthFirstSearch(Configuration configuration) {
        conf = configuration;
    }
```

3. 下面的 run() 方法设置 Giraph 作业的配置项:

```java
    @Override
    public int run(String[] args) throws Exception {
        if(args.length < 5) {
            System.err.println(printUsage());
            System.exit(1);
        }
        if(args.length > 5) {
            System.err.println("too many arguments. " +"Did you forget to quote the source or destination ID name ('firstname lastname')");
            System.exit(1);
        }
        String input = args[0];
        String output = args[1];
        String source_id = args[2];
        String dest_id = args[3];
        String zooQuorum = args[4];

        conf.set(SOURCE_ID, source_id);
        conf.set(DEST_ID, dest_id);
        conf.setBoolean(GiraphJob.SPLIT_MASTER_WORKER, false);
        conf.setBoolean(GiraphJob.USE_SUPERSTEP_COUNTERS, false);
        conf.setInt(GiraphJob.CHECKPOINT_FREQUENCY, 0);
        GiraphJob job = new GiraphJob(conf, "determine connectivity between" + source_id + " and " + dest_id);
        job.setVertexClass(EmployeeSearchVertex.class);
        job.setVertexInputFormatClass(EmployeeRDFTextInputFormat.class);
        job.setVertexOutputFormatClass(BreadthFirstTextOutputFormat.class);
        job.setZooKeeperConfiguration(zooQuorum);
        FileInputFormat.addInputPath(job.getInternalJob(), new Path(input));
        FileOutputFormat.setOutputPath(job.getInternalJob(),   removeAndSetOutput(output))

        job.setWorkerConfiguration(1, 1, 100.0f);

        if(job.run(true)) {
            long srcCounter = job.getInternalJob().getCounters().
                    getGroup("Search").findCounter("Source Id found").getValue();
            long dstCounter =
```

```
job.getInternalJob().getCounters().getGroup("Search").findCounter("Dest
Id found").getValue();
            if(srcCounter == 0 || dstCounter == 0) {
                System.out.println("Source and/or Dest Id not found in
dataset. Check your arguments.");
            }
            return 0;
        } else {
            return 1;
        }
    }
```

4. 如下的代码,强行删除已存在 HDFS 的输出目录。注意小心使用。其他方法需要和 Tool 接口一致:

```
    private Path removeAndSetOutput(String outputDir) throws
IOException {
        FileSystem fs = FileSystem.get(conf);
        Path path = new Path(outputDir);
        fs.delete(path, true);
        return path;
    }
    private String printUsage() {
        return "usage: <input> <output> <single quoted source_id>
<single quoted dest_id> <zookeeper_quorum>";
    }

    @Override
    public void setConf(Configuration conf) {
        this.conf = conf;
    }

    @Override
    public Configuration getConf() {
        return conf;
    }
```

5. main() 方法使用 ToolRunner 实例化并提交该作业:

```
    public static void main(String[] args) throws Exception {
        System.exit(ToolRunner.run(new EmployeeBreadthFirstSearch (new
Configuration()), args));
    }
```

6. 静态内部类 EmployeeBreadthFirstSearchVertex 为每个大步过程定义了一个用户自定义的计算方法:

```
    public static class EmployeeSearchVertex<I extends WritableComparable,
V extends Writable, E extends Writable, M extends Writable> extends
EdgeListVertex<Text, IntWritable, NullWritable, IntWritable> {
```

```
            private IntWritable msg = new IntWritable(1);

            private boolean isSource() {
                return getId().toString().equals(
                        getConf().get(SOURCE_ID));
            }

            private boolean isDest() {
                return getId().toString().equals(
                        getConf().get(DEST_ID));
            }

            @Override
            public void compute(Iterable<IntWritable> messages) throws IOException {
                if(getSuperstep() == 0) {
                    if(isSource()) {
                        getContext().getCounter("Search", "Source Id found").
increment(1);
                        sendMessageToAllEdges(msg);
                    } else if(isDest()){
                        getContext().getCounter("Search", "Dest Id found").
increment(1l);
                    }
                }
                boolean connectedToSourceId = false;
                for(IntWritable msg : messages) {
                    if(isDest()) {
                        setValue(msg);
                    }
                    connectedToSourceId = true;
                }
                if(connectedToSourceId)
                    sendMessageToAllEdges(msg);
                voteToHalt();
            }
        }
```

7. 内部静态类 `BreadthFirstTextOutputFormat` 定义了用户自定义 `OutputFormat`。类 `BreadthFirstTextOutputFormat` 按照 `Text` 类型的键值对输出节点信息到 HDFS 中。

```
        public static class BreadthFirstTextOutputFormat extends TextVertexOutputFormat
<Text, IntWritable, NullWritable> {

            private static class EmployeeRDFVertexWriter extends TextVertexWriter
<Text, IntWritable, NullWritable> {

                private Text valOut = new Text();
                private String sourceId = null;
                private String destId = null;
```

```
            public EmployeeRDFVertexWriter(String sourceId, String destId,
RecordWriter<Text, Text> lineRecordWriter) {
                super(lineRecordWriter);
                this.sourceId = sourceId;
                this.destId = destId;
            }

            @Override
            public void writeVertex(
                Vertex<Text, IntWritable, NullWritable, ?> vertex)
                    throws IOException, InterruptedException {
                        if(vertex.getId().toString().equals(destId)) {
                    if(vertex.getValue().get() > 0) {
                        getRecordWriter().write(new Text(sourceId + " is
connected to " + destId), new Text(""));
                    } else {
                        getRecordWriter().write(new Text(sourceId + " is
not connected to " + destId), new Text(""));
                    }
                }
            }
        }

        @Override
        public VertexWriter<Text, IntWritable, NullWritable>
        createVertexWriter(TaskAttemptContext context) throws IOException,
InterruptedException {
            RecordWriter<Text, Text> recordWriter =
                textOutputFormat.getRecordWriter(context);
            String sourceId = context.getConfiguration().get(SOURCE_ID);
            String destId = context.getConfiguration().get(DEST_ID);
            return new EmployeeRDFVertexWriter(sourceId, destId, recordWriter);
        }
    }
}
```

8. 创建 shell 脚本 run_employee_connectivity_search.sh，其中的代码命令如下。将 GIRAPH_PATH 修改为 Giraph JAR 文件的本地路径，将 JAR_PATH 修改为刚才编译的自定义 JAR 文件的本地路径。

 为了在自定义 JAR 文件中使用别名 emp_breadth_first，必须在主类中使用 Hadoop 提供的 Driver 类。

```
GIRAPH_PATH=lib/giraph/giraph-0.2-SNAPSHOT-jar-with-dependencies.jar
HADOOP_CLASSPATH=$HADOOP_CLASSPATH:$GIRAPH_PATH
JAR_PATH=dist/employee_examples.jar
export HADOOP_CLASSPATH
hadoop jar $JAR_PATH emp_breadth_search -libjars $GIRAPH_PATH, $JAR_PATH
/input/gooftech /output/gooftech 'Valery Dorado' 'Gertha Linda' localhost:2181
```

9. 执行 `run_employee_connectivity_search.sh`。这个作业将会被提交至 Hadoop 集群。在作业结束之后，路径 `/output/gooftech` 会得到一个单 part 的文件，里面内容是 `Valery Dorado is not connected to Gertha Linda`。

10. 打开 `run_employee_connectivity_search.sh`，修改源 ID 为 `Shoshana Gatton`。保存并关闭脚本。

11. 执行 `run_employee_connectivity_search.sh`，输出结果将是 `Shoshana Gatton is connected to Gertha Linda`。

工作原理

如果希望了解自定义 `InputFormat` 以及作业的初始化是如何进行的，可以参见 7.3 节的 "工作原理" 部分的内容。本节使用相同的输入格式、相同的作业初始化过程，除了下面的一些方面存在不同。

- 作业需要额外的 `DEST_ID` 参数提供给命令行。
- 节点的实现类为 `EmployeeSearchVertex`。
- `OutputFormat` 的子类 `BreadthFirstTextOutputFormat` 设置为静态内部类。更多的解释会在下面的篇章中出现。
- 在作业的执行过程中，使用计数器来确定提供的源/目标 ID 是否在数据集中出现。

类 `EmployeeSearchVertex` 中的方法 `Compute()` 通过 Giraph 的消息传递确定可达性。在 S0 的开始阶段，从源 ID 发送一个消息至与之相连的每条边。如果发现提供的源 ID 与目标 ID 出现在数据集节点中，将增加对应的计数器，使用户知道相应信息。这将帮助我们快速地知道在命令行参数中是否输入任何不正确的源/目标节点 ID。在 S0 阶段之后，这两个计数器将会被置为 1。定义了一个私有常量成员变量 `msg`，设置为 1。实际的数值类型的消息内容未被使用，但为了使用已经创建的自定义 `InputFormat` `EmployeeRDFTextInputFormat`，保持节点值的数据类型为 `IntWritable`。在之后的大步带有节点接收到一条消息，将会查看消息传递经过的每个边。如果目标节点已经收到一个消息，会设置消息包含整数 1。在作业执行结束时，目标节点会获得一个值 "1"，表明目标节点通过一条或多条边连接至源节点。如果仍然为初始值 "0"，表明没有收到任何消息，这时是非连通的。

定义的静态内部类 `BreadthFirstTextOutputFormat` 处理输出格式。这个类与我们之前自定义的 `InputFormat` 的继承/委托模式类似，不同的是委托模式使用自定义类 `RecordWriter` 代替自定义类 `RecordReader`。我们实例化了 `TextVertexWriter` 的子类 `EmployeeRDFVertexWriter`，通过引用配置源/目标 ID。框架自动处理包含在数据集中的

每个节点的方法 `writeVertex()` 的调用。该作业只关心打印出是否源节点可通过一个或多个路径连接上目标节点。如果处理的当前节点是目标节点，将输出两种可能字符串的其中一个。如果节点值大于 0，那么目标节点一定收到一个或多个消息，表明至少存在一条路径连接源节点与目标节点。否则，如果目标节点仍然为 0，可以大胆地推断目标节点对于源节点是不可达的。对于这样一对源-目标节点，根据上面的代码，能在作业执行完成后利用计数器在作业类中直接替换相关业务逻辑。但如果希望使用上述代码查询多个源-目标节点对，需要对这个设计进行一些拓展。

更多参考

当开始尝试在大规模数据范围内测试时，使用 Hadoop MapReduce API 设计程序，需要额外的一些调试。对于不能简单扩展至大规模数据的程序，通常需要完整的重新评估，选择设计更高的程序模式。使用 Giraph API 同样需要多多尝试并耐心调试。

Apache Giraph 作业通常需要可扩展性的调试

在开始的时候通常不容易发现。对于相对较小的图，使用上面给出的 BSP 方法，程序能很好的运行。当突然遇到扩展性的问题时，会发现遇到意料之外的各种错误。确保 `compute()` 函数尽量简单，避免函数变得复杂，从而可以避免发生故障。同时，Giraph 的 worker 会尝试直接在内存保存分配给它的图划分。最小化节点的内存占用是非常重要的。另外，很多人需要通过设置位于 `GiraphJob` 的参数，调整传递的消息。可以通过设置参数 `MSG_NUM_FLUSH_THREADS`，控制每个 worker 连接其他 worker 的消息线程数。默认的，Giraph 会在作业中为每个 worker 开启一个线程与其他 worker 通信。对于大多数 Hadoop 集群，这是不可忍受的。同样可以考虑使用参数 `MAX_MESSAGES_PER_FLUSH_PUT` 调整允许消息一次写回磁盘的最大数。默认参数 2000 可能不适合你的作业。

7.5 使用 Apache Mahout 计算协同过滤

协同过滤是一种用于发现用户与商品项(比如，书和音乐)之间关系的技术。通过分析用户的喜好集合，例如用户购买过的商品项，确定用户之间共同的偏好，从而获得用户与商品之间的关系。协同过滤可以用于创建推荐系统。推荐系统被许多互联网公司使用，例如：Amazon、LinkIn 以及 FaceBook。

本节基于包含用户数据偏好的数据集，使用 Apache Mahout，创建一个图书推荐系统。

准备工作

需要下载、编译并安装如下内容。

- 从 http://maven.apache.org/ 下载 Maven 2.2 或更高版本。
- 从 http://mahout.apache.org/ 下载 Apache Mahout 0.6。
- 从 http://www.informatik.uni-freiburg.de/~cziegler/BX/ 下载 CSV 格式的 Book-Crossing 数据集。
- 从 http://packtpub.com/support 下载本章的脚本。

一旦编译 Mahout 完成，可以将 mahout 二进制文件添加至系统路径。此外，需要设置环境变量 HADOOP_HOME 指向安装 Hadoop 的根文件夹。通过下面的命令行在 bash shell 中完成上面的操作：

```
$ export PATH=$PATH:/path/to/mahout/bin
$ export HADOOP_HOME=/opt/mapr/hadoop/hadoop-0.20.2
```

接下来，解压 Book-Crossing 数据集至当前的工作文件夹。可以看到包含的三个文件，分别是 BX-Books.csv、BX-Book-Ratings.csv 和 BX-Users.csv。

操作步骤

执行下面的步骤，使用 Mahout 执行协同过滤。

1. 运行脚本 clean_book_ratings.py 将文件 BX-Book-Ratings.csv 转化为 Mahout 推荐系统能使用的格式。

```
$ ./clean_book_ratings.py BX-Book-Ratings.csv cleaned_book_ratings.txt
```

2. 运行 bash 脚本 clean_book_users.sh 将文件 BX-Users.csv 转化为 Mahout 推荐系统能使用的格式。注意：文件 BX-Users.csv 需要保存在当前工作路径下。

```
$ ./clean_book_users.sh
```

3. 加载文件 cleaned_book_ratings.txt 以及 cleaned_book_users.txt 至 HDFS。

```
$ hadoop fs -mkdir /user/hadoop/books
$ hadoop fs -put cleaned_book_ratings.txt /user/hadoop/books
$ hadoop fs -put cleaned_book_users.txt /user/hadoop/books
```

4. 使用刚才放置在 HDFS 的用户打分以及用户信息文件运行 Mahout 推荐系统。Mahout 会启动多个 MapReduce 作业生成图书的推荐结果：

```
$ mahout recommenditembased --input /user/hadoop/books/ cleaned_
```

```
book_ratings.txt --output /user/hadoop/books/recommended
--usersFile /user/hadoop/books/cleaned_book_users.txt -s
SIMILARITY_LOG_LIKELIHOOD
```

5. 检查结果的格式：`USERID [RECOMMENDED BOOK ISBN:SCORE,…]`。输出的结果应该与下面的类似：

```
$ hadoop fs -cat /user/hadoop/books/recommended/part* | head -n1
17      [849911788:4.497727,807503193:4.497536,881030392:4.497536,
761528547:4.497536,380724723:4.497536,807533424:4.497536,310203414:4.497
536,590344153:4.497536,761536744:4.497536,531000265:4.497536]
```

6. 使用 `print_user_summaries.py` 以更人性化的方式检查结果。打印最前面 10 个用户的推荐结果。这里使用 10 作为 `print_user_summaries.py` 的最后一个参数：

```
hadoop fs -cat /user/hadoop/books/recommended/part-r-00000 | ./
print_user_summaries.py BX-Books.csv BX-Users.csv BX-Book-Rating.
csv 10
==========
user id =  114073
rated:
Digital Fortress : A Thriller  with:  9

Angels &amp Demons with:  10

recommended:
Morality for Beautiful Girls (No.1 Ladies Detective Agency)
Q Is for Quarry
The Last Juror
The Da Vinci Code
Deception Point
A Walk in the Woods: Rediscovering America on the Appalachian
Trail (Official Guides to the Appalachian Trail)
Tears of the Giraffe (No.1 Ladies Detective Agency)
The No. 1 Ladies' Detective Agency (Today Show Book Club #8)
```

`print_user_summaries.py` 输出了用户关于书的评分以及通过 Mahout 生成的推荐结果。

工作原理

第一步需要我们清洗 Book-Crossing 数据集。文件 `BX-Book-Ratings.csv` 是以分号分隔的，文件的列包括如下内容。

- `USER_ID`：每个人分配的 ID 值。
- `ISBN`：每个人看过的书的 ISBN。
- `BOOK-RATING`：用户给书打过的分。

Mahout 推荐引擎希望数据集以逗号分隔格式输入下面的内容。

- `USER_ID`：USER_ID 必须是一个整数。
- `ITEM_ID`：ITEM_ID 必须是一个整数。
- `RATING`：RATING 必须是描述喜爱程度的整数。例如，1 表示用户非常不喜欢这本书，10 表示用户非常喜爱这本书。

一旦完成文件 `BX-Book-Ratings.csv` 的转换，继续对文件 `BX-Users.csv` 执行类似的转换。除了 `USER_ID` 之外，这里略去了文件 `BX-Users.csv` 的信息。

最后，启动 Mahout 推荐引擎。Mahout 会启动一系列 MapReduce 作业。通过 `-usersFile` 标志位指定一个用户集合，Mahout 为这个用户集合进行书籍推荐。在这个例子中，我们希望 Mahout 生成数据集中所有用户的推荐结果，因此我们提供完整的 `USER_ID` 列表给 Mahout。除了以 Mahout 参数的形式提供输入路径、输出路径以及用户列表，我们还指定了第四个参数 `-s SIMILARITY_LOGLIKELIHOOD`。标志 `-s` 指定 Mahout 使用的相似度测量方式，比较所有用户关于书的喜爱程度。本节使用对数似然估计是因为这是一个简单而有效的算法，同时 Mahout 提供了更多的相似度函数。需要了解更多内容，你可以运行下面的命令，检查标志 `-s` 的可选项：

```
$mahout recommenditembased
```

延伸阅读

- 使用 Apache Mahout 进行聚类（7.6 节）。
- 使用 Apache Mahout 进行情感分类（7.7 节）。

7.6 使用 Apache Mahout 进行聚类

聚类是一种将数据集划分为若干相关子集的技术。本节我们使用特定的聚类方法：k-means。k-means 聚类尝试划分数据集至 k 个聚类，根据聚类中各点到中心点的最小距离进行聚类。

具体而言，本节使用 Apache Mahout 的 k-means 方法对莎士比亚的悲剧作品集中出现的词进行聚类。

准备工作

需要下载、编译并安装如下内容。

- 从 `http://maven.apache.org/` 下载 Maven 2.2 或更高版本。
- 从 `http://mahout.apache.org/` 下载 Apache Mahout 0.6。
- 从 `http://packtpub.com/support` 下载文件 `Shakespeare.zip`。

解压 `Shakespeare.zip` 至文件夹 `shakespeare_text`。`Shakespeare.zip` 压缩文件包含莎士比亚的六部作品。将 `shakespeare_text` 文件夹与其中的内容放置在 HDFS。

```
$ mkdir shakespeare_text
$ cd shakespeare_text
$ unzip shakespeare.zip
$ cd ..
$ hadoop fs -put shakespeare_text /user/Hadoop
```

操作步骤

执行下面的步骤，使用 Mahout 完成聚类。

1. 转换莎士比亚文章为 Hadoop 的 SequenceFile 格式：

```
mahout seqdirectory --input /user/hadoop/shakespeare_text --output /user/hadoop/shakespeare-seqdir --charset utf-8
```

2. 转换包含文本内容的 SequenceFile 文件为一个向量：

```
mahout seq2sparse --input /user/hadoop/shakespeare-seqdir
--output /user/hadoop/shakespeare-sparse --namedVector -ml 80
-ng 2 -x 70 -md 1 -s 5 -wt tfidf -a org.apache.lucene.analysis.
WhitespaceAnalyzer
```

3. 在文本向量上运行 k-means 聚类算法。该命令行会启动 10 个 MapReduce 作业。同样的由于使用 k-means 聚类，需要指定类别个数：

```
mahout kmeans --input /user/hadoop/shakespeare-sparse/tfidf-
vectors --output /user/hadoop/shakespeare-kmeans/clusters
--clusters /user/hadoop/shakespeare-kmeans/initialclusters
--maxIter 10 --numClusters 6 --clustering -overwrite
```

4. 使用下面的命令行检查 Mahout 的聚类标示：

```
mahout clusterdump --seqFileDir /user/hadoop/shakespeare-kmeans/
clusters/clusters-1-final --numWords 5 --dictionary /user/hadoop/
shakespeare-sparse/dictionary.file-0 --dictionaryType sequencefile
```

工具 `clusterdump` 的输出结果是十分庞大的。可以查看输出的高频词部分。例如，下面是通过 k-means 算法识别的《罗密欧与朱丽叶》类别的高频词：

```
r=/romeoandjuliet.txt =]}
        Top Terms:
                ROMEO                          =>   29.15485382080078
```

```
            JULIET                          =>      25.78818130493164
            CAPULET                         =>      21.401729583740234
            the                             =>      20.942245483398438
            Nurse                           =>      20.129182815551758
```

工作原理

开始步骤需要我们基于原始数据做一些进行 Mahout 的 k-means 算法的前期预处理工作。工具 `seqdirectory` 简单地转换文件夹中的相关内容至 SequenceFile 文件。接下来，工具 `seq2sparse` 转化新生成的 SequenceFile 文件（仍然包含文本内容）为文本向量。`seq2sparse` 的参数描述如下。

- `--input`：Mahout 所需的包含 SequenceFile 文件格式的 HDFS 文件夹。

- `--output`：存储文本向量的 HDFS 输出路径。

- `--namedVector`：用于命名向量的标志。

- `-ml`：最小对数自然估计阈值。我们设置相对高的数字，因为我们需要得到最明显的词组聚类。

- `-ng`：n-gram 的大小。

- `-x`：一个阈值，规定了一个词条在文章中出现的频率超过多大百分比时，它就应该被忽略。本节我们选择 70，意味着任何出现频率超过 70% 文档的词条都将会被忽略。使用该设置可去除掉意义不大的词（例如，词：at、a 以及 the）。

- `-md`：一个词条在被考虑处理之前，需要出现在所有文本的最少次数。本节，我们使用 1，表示在文章中出现过至少一次的词条将会被处理。

- `-s`：一个词条在被考虑处理之前，需要出现在一篇文档的最少次数。

- `-wt`：需要使用的权重算法。这里我们选择使用 TF-IDF。另外的选项是 TF，但这种方法不能有效地识别重要的 n-gram。

- `-a`：需要使用的分析器的种类。分析器用于转化文本文档。WhitespaceAnalyzer 使用空格分割文档为单词标示（Token）。单词标示会根据应用 `seq2sparse` 的其他标志位进行保留、合并或者忽略操作。

最后，我们在莎士比亚数据集上运行 k-means 聚类算法。Mahout 会启动一系列 MapReduce 作业，这里的作业是可以配置的。如果 k-means 聚类收敛了，或者达到了 MapReduce 作业最大数，k-means 作业将会完成。下面是一些允许配置的 k-means Mahout 作业参数的定义。

高级大数据分析 171

- `--input`：包含文本向量的 HDFS 文件夹。
- `--output`：k-means 作业的 HDFS 输出路径。
- `--maxIter`：启动 MapReduce 作业的最大数。
- `--numClusters`：需要识别的聚类数目。我们选择 6 因为数据集是六部莎士比亚的作品，我们希望识别出这些文章明显的 bigram。
- `--clusters`：类别的初始化点信息。
- `--clustering`：告知 Mahout 在聚类之前需要先遍历一遍数据的标志位。
- `--overwrite`：告诉 Mahout 输出文件夹的标志位。

延伸阅读

- 使用 Apache Mahout 进行情感分类（7.6 节）。

7.7 使用 Apache Mahout 进行情感分类

情感分类是一种尝试确定一个人对一类商品项是否有喜爱倾向的分类处理过程。本节，我们使用 Apache Mahout 提供的朴素贝叶斯分类器确定对电影评价的一个词组集合是否对该影片有正面或负面的态度。

准备工作

需要下载、编译并安装如下内容。

- 从 `http://maven.apache.org/` 下载 Maven 2.2 或更高版本。
- 从 `http://mahout.apache.org/` 下载 Apache Mahout 0.6。
- 从 `http://www.cs.cornell.edu/people/pabo/movie-review-data/` 下载 `Polarity_dataset_v2.0`。
- 从 `http://packtpub.com/support` 下载本节的脚本。

在当前工作路径解压电影评价数据集 `review_polarity.tar.gz`，将得到名为 `txt_sentoken` 的文件夹。在文件夹中另外包含两个文件夹，分别为 `pos` 以及 `neg`。文件夹 `pos` 以及 `neg` 包含电影评价的文本文件。显而易见的，`pos` 文件夹包含正面的电影评论，而 `neg` 文件夹包含负面的电影评价。

操作步骤

1. 在当前工作目录运行 reorg_data.py 脚本，为 Mahout 分类器转换数据成训练集和测试集：

```
$ ./reorg_data.py txt_sentoken train test
```

2. 为 Mahout 分类器准备相应数据集：

```
$ mahout prepare20newsgroups -p train -o train_formated -a org.apache.mahout.vectorizer.DefaultAnalyzer -c UTF-8
$ mahout prepare20newsgroups -p test -o test_formated -a org.apache.mahout.vectorizer.DefaultAnalyzer -c UTF-8
```

这个例子读写本地文件系统而非 HDFS。

3. 放置 train_formated、test_formated 至 HDFS：

```
$ hadoop fs -put train_formated /user/hadoop/
$ hadoop fs -put test_formated /user/hadoop/
```

4. 使用 train_formated 数据集训练朴素贝叶斯分类器：

```
$ mahout trainclassifier -i /user/hadoop/train_formated -o /user/hadoop/reviews/naive-bayes-model -type bayes -ng 2 -source hdfs
```

5. 使用 test_formated 数据集测试分类器：

```
$ mahout testclassifier -m /user/hadoop/reviews/naive-bayes-model -d prepared-test -type bayes -ng 2 -source hdfs -method sequential
```

6. 工具 testclassifier 返回相似度总结以及混淆矩阵。具体的数可能与下面显示的数字不相同：

```
Summary
-------------------------------------------------------
Correctly Classified Instances      :     285      71.25%
Incorrectly Classified Instances    :     115      28.75%
Total Classified Instances          :     400
=======================================================
Confusion Matrix
-------------------------------------------------------
a      b       <--Classified as
97     103     |  200         a      = pos
12     188     |  200         b      = neg
```

工作原理

开始的两步需要为 Mahout 朴素贝叶斯分类器准备数据。脚本 reorg_data.py 从 txt_sentoken 文件夹分发正负面评论至训练集和测试集。80%的评论会放置于训练集，剩下 20%的评论作为测试集。接下来，使用工具 `prepare20newsgroups` 转换训练集和测试集的格式为 Mahout 分类器兼容的格式。Mahout 中的例子数据集与脚本 `reorg_data.py` 生成的数据有相似的格式，因此我们可以使用工具 `prepare20newsgroups`。`prepare20newsgroups` 所做的事包括根据数据集的分类（正面或负面数据），合并所有 `pos`、`neg` 文件夹中的文件至一个文件。这样，替代了 1000 个只包含一条正面或负面评论的文件，取而代之的是，拥有所有正面或负面评论的两个文件 `pos.txt` 以及 `neg.txt`。

接下来，使用 n-gram 大小为 2，指定了 `-ng` 标志位，在 HDFS 上的 `train_formated` 数据集，训练朴素贝叶斯分类器。Mahout 通过启动一系列 MapReduce 作业训练分类器。

最后，运行 `testclassifier` 工具，使用 HDFS 上的 `test_formated` 数据，测试在步骤 4 创建的分类器。在步骤 6 会看到，对于测试数据集分类的正确率为 71.25%。这里需要注意的是，这个统计并不代表分类器对每一个电影评论有 71.25%的正确性。这里有多种训练并验证分类器的方法，这些技术超出了本书这一章的讨论范围。

更多参考

第 6 步使用的工具 `testclassifier` 并非运行 MapReduce 作业，这是在用本地模式测试分类器。如果需要使用 MapReduce 测试分类器，可以修改 `-method` 参数为 `mapreduce`。

```
$ mahout testclassifier -m /user/hadoop/reviews/naive-bayes-model -d
prepared-test -type bayes -ng 2 -source hdfs -method mapreduce
```

第 8 章

调试

本章我们将介绍：
- 在 MapReduce 中使用 Counters 监测异常记录
- 使用 MRUnit 开发和测试 MapReduce
- 本地模式下开发和测试 MapReduce
- 运行 MapReduce 作业跳过异常记录
- 在流计算作业中使用 Counters
- 更改任务状态显示调试信息
- 使用 illustrate 调试 Pig 作业

8.1 介绍

在 Hadoop 中有这么一句格言："everything breaks at scale"。畸形和异常的输入数据是十分常见的。处理大量的异常数据本身就是一个很大的问题。在 Hadoop 的上下文中，不同的任务是相互独立的，各自处理着不同的输入数据，这使得 Hadoop 能够很容易地分发作业，却很难监测全局事件和理解每个任务的状态。幸运的是，有一些工具和技术可以帮助调试 Hadoop 作业。本章我们将重点介绍如何使用这些工具和技术调试 MapReduce 作业。

8.2 在 MapReduce 中使用 Counters 监测异常记录

在 MapReduce 框架中提供了 **Counters** 这个可以在 map 阶段和 reduce 阶段有效监测全

局事件的机制。举个例子，一个经典的 MapReduce 作业会启动多个 mapper 实例，每个输入数据块上都将运行同样的代码。这些实例是同一个作业的一部分，但是彼此又相互独立。Counters 允许开发人员监测并聚合这些独立实例发生的事件。

Counters 的一个更具体的使用可以在 MapReduce 自身框架找到。每个 MapReduce 作业都定义了一些独立的计数器。这些计数器的输出可以在 Job Tracker Web 界面的作业详细信息中看到。

	Counter	Map	Reduce	Total
Map-Reduce Framework	Map input records	71,228,085,554	0	71,228,085,554
	Reduce shuffle bytes	0	5,419,119,866,175	5,419,119,866,175
	Spilled Records	142,174,003,132	55,141,286,320	197,315,289,452
	Map output bytes	5,513,542,463,958	0	5,513,542,463,958
	CPU_MILLISECONDS	2,658,192,290	2,352,883,960	5,011,076,250
	Combine input records	139,937,676,997	69,478,386,276	209,416,063,273
	SPLIT_RAW_BYTES	1,999,232	0	1,999,232
	Reduce input records	0	58,006,566,989	58,006,566,989
	Reduce input groups	0	3,331,430	3,331,430
	Combine output records	139,937,676,997	69,478,386,262	209,416,063,259
	PHYSICAL_MEMORY_BYTES	12,647,158,763,520	196,662,726,656	12,843,821,490,176
	Reduce output records	0	406,433,728	406,433,728
	VIRTUAL_MEMORY_BYTES	33,014,737,301,504	273,806,381,056	33,288,543,682,560
	Map output records	71,228,085,554	0	71,228,085,554
	GC time elapsed (ms)	98,623,849	342,650,729	441,274,578

UI 界面显示了计数器的组、名称、mapper 统计、reducer 统计和作业统计。

Counters 应当只限于跟踪作业相关的元数据。基础计数器就是一个很好的例子。Map 输入记录计数器提供了关于一个特定作业执行的有用信息。如果没有计数器的存在，那么这些类型的统计务必成为作业输出的一部分，或者说是输出的另一部分，但这些信息本不属于作业的输出，会使作业的逻辑复杂化。

接下来将介绍一个简单的只有 map 的作业，过滤出日志中的异常记录，并使用计数器对异常数据的数量进行统计。

准备工作

从 http://www.packtpub.com/support 这个网址下载 weblog_entries_bad_records.txt 数据集。

操作步骤

1. 将 weblog_entries_bad_records.txt 文件上传到 HDFS 上的一个新目录：
`hadoop fs -copyFromLocal weblog_entries.txt /data/weblogs`

2. 提交 `CountersExample` 作业：

`hadoop jar ./CountersExample.jar com.packt.hadoop.solutions.CounterExample /data/weblogs/weblog_entries_bad_records.txt /data/weblogs/weblog_entries_clean.txt`

3. 打开浏览器通过 Job Tracker 的 UI 界面查看作业的计数器信息。默认的地址是 `localhost:50030`。向下滚动到 `Completed Jobs` 段，然后定位到 **CounterExample** 作业。最新的作业排在列表的最底部，一旦定位到作业就可以点击 Jobid。该页有关于统计的一些高级信息，包含计数器信息。

com.packt.hadoop.solutions.CounterExample$BadRecords	INVALID_NUMBER_OF_COLUMNS	2	0	2
	INVALID_IP_ADDRESS	2	0	2

工作原理

计数器是按照组来进行定义的。在 Java 中每个计数器分组是一个枚举类型。在 CounterExample 作业中，监测异常数据计数的定义如下：

`static enum BadRecords{INVALID_NUMBER_OF_COLUMNS, INVALID_IP_ADDRESS};`

在 map 函数中，存在两处数据有效性监测。第一次检查将数据按制表符进行分隔。在这个例子中如果数据是合法的，那么每一条记录应该都包含 5 个列。如果某条记录不是包含 5 个列，就会调用 `Context` 类中的 `BadRecords.INVALID_NUMBER_OF_COLUMNS` 计数器。该计数器值将加 1。

```
String record = value.toString();
String [] columns = record.split("\t");

// Check for valid number of columns
if (columns.length != 5) {
context.getCounter(BadRecords.INVALID_NUMBER_OF_COLUMNS).increment(1);
return;
}
```

第二次是检查 IP 地址的有效性。定义了一个 `VALID_IP_ADDRESS` 的正则表达式。从名称就可以看出，这个正则表达式将匹配有效的 IP 地址。

```
private static final String VALID_IP_ADDRESS = "^([01]?\\d\\d?|2[0-4]\\d|25[0-5])\\.([01]?\\d\\d?|2[0-4]\\d|25[0-5])\\." +
    "([01]?\\d\\d?|2[0-4]\\d|25[0-5])\\.([01]?\\d\\d?|2[0-4]\\ d|25[0-5])$";
```

`VALID_IP_ADDRESS` 正则表达式用于监测每一条记录的 IP 地址是否匹配。任何一条记录不匹配 `INVALID_IP_ADDRESS` 计数器将自增。

```
// Check for valid IP addresses
Matcher matcher = pattern.matcher(columns[4]);
If (!matcher.matches()) {
    context.getCounter(BadRecords.INVALID_IP_ADDRESS).increment(1);
    return;
}
```

每个自增的计数器都先在各自 mapper 本地存储。计数器的统计值每隔一秒会发送到 Task Tracker 进行聚合，最后这些统计值会发送到 Job Tracker 进行全局聚合。

8.3 使用 MRUnit 开发和测试 MapReduce

从概念上讲，MapReduce 作业运行机制相对简单。在 map 阶段每一条输入记录都会进行一定处理，生成一个或者多个键值对。在 reduce 阶段按组接收这些键值对，并且进行一定的逻辑处理。测试 Mappers 和 Reducers 函数应当与测试其他函数一样简单。一个给定的输入会产生可预期的输出。复杂的地方在于 Hadoop 的分布式特性。Hadoop 是一个包含很多可移动部件的大框架。在 Cloudera 发布 MRUnit 之前，即使在本地模式下最简单的测试也必须从磁盘上读取数据并至少需要几秒进行初始化和运行。

在开发和测试的时候，MRUnit 尽可能消除了 Hadoop 的框架性。重点是减少 map 代码和 reduce 代码，以及输入数据与期望输出的结果。使用 MRUnit，开发和测试 MapReduce 代码完全可以在 IDE 上完成，并且可以在一瞬间完成测试。

本节将演示 MRUnit 如何使用在 `lib` 文件夹中 MapReduce 框架提供的 IdentityMapper。该 IdentityMapper 以键值对作为输入并输出不变的键值对。

准备工作

按照如下步骤开始。

- 从 http://mrunit.apache.org/general/downloads.html 下载 MRUnit 的最新版本。
- 创建一个 Java 工程。
- 将 `mrunit-X.Y.Z-incubating-hadoop1.jar` 以及其他 Hadoop 的 JAR 文件添加到 Java 工程的构建路径中。
- 创建一个类，命名为 `IdentityMapperTest`。
- 在本章的源代码文件夹中查看 `IdentityMapperTest.java` 的完整源代码。

操作步骤

按照以下几个步骤实现使用 MRUnit 对 mapper 函数进行测试。

1. 使 `IdentityMapperTest` 继承 `TestCase` 类：

```
public class IdentityMapperTest extends TestCase
```

2. 创建两个私有成员变量 mapper 和 driver：

```
private Mapper identityMapper;
private MapDriver mapDriver;
```

3. 在 `setup()` 方法上添加 `Before` 注释：

```
@Before
public void setup() {
    identityMapper = new IdentityMapper();
    mapDriver = new MapDriver(identityMapper);
}
```

4. 在 `testIdentityMapper1()` 方法上添加 `Test` 注释：

```
@Test
public void testIdentityMapper1() {
    mapDriver.withInput(new Text("key"), new Text("value"))
    mapDriver.withOutput(new Text("key"), new Text("value"))
            .runTest();
}
```

5. 运行测试程序。

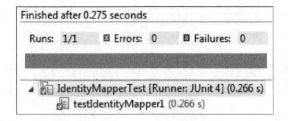

6. 添加一个错误的测试用例 `testIdentityMapper2`：

```
@Test
public void testIdentityMapper2() {
    mapDriver.withInput(new Text("key"), new Text("value"))
    mapDriver.withOutput(new Text("key2"), new Text("value2"))
```

```
            mapDriver.runTest();
};
```

7. 再运行一次。

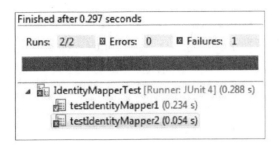

工作原理

MRUnit 是构建在流行的 Junit 测试框架之上的。它使用了模拟对象的类库——Mockito，模拟了 Hadoop 大部分重要的类，这样用户只需要关注 map 和 reduce 的逻辑。`MapDriver` 是运行测试的驱动类，是 `Mapper` 类的一个实例化。调用 `WithOutput()` 方法提供用来校验 map 输出结果的正确性。调用方法 `runTest()` 实际上是调用 mapper 并传递输入数据，再通过 `withOutput` 方法校验输出结果。

更多参考

这个例子只是展示了如何对 mapper 函数进行测试。MRUnit 还提供了一个 `ReduceDriver` 类，可以使用与 `MapDriver` 相同的方式对 reduce 进行测试。

延伸阅读

- 更多关于 Mockito 的信息可以访问 http://code.google.com/p/mockito/。
- 本地模式下开发和测试 MapReduce（8.4 节）。

8.4 本地模式下开发和测试 MapReduce

利用 MRUnit 开发和本地模式开发两种方式是互补的。MRUnit 提供了一种优雅的测试方式用于测试 MapReduce 作业 map 与 reduce 两个阶段。最初的开发和测试工作应该使用这个框架。但是，运行 MRUnit 测试时，MapReduce 作业的几个关键组件并没有运用到。包括两个关键的类 `InputFormats` 和 `OutFormats`。在本地模式下运行作业将测试得更全面。本地模式下，测试时也更容易使用大量的真实数据。

本节将展示如何配置 hadoop 运行在本地模式，并且使用 Eclipse 的调试器组件进行调试。

准备工作

你需要从 Packet 的网站 http://www.packtpub.com/support 下载数据集 weblog_entries_bad_records.txt。这个例子将用在 8.2 节中提供的 CounterExample.java。

操作步骤

1. 在文本编辑器中打开 $HADOOP_HOME/conf/mapred-site.xml 文件。

2. 设置 mapred.job.tracker 属性值为 local：

```
<property>
    <name>mapred.job.tracker</name>
    <value>local</value>
</property>
```

3. 在文本编辑器中打开 $HADOOP_HOME/conf/core-site.xml 文件。

4. 设置 fs.default.name 属性值为 file:///。

```
<property>
  <name>fs.default.name</name>
  <value>file:///</value>
</property>
```

5. 打开 $HADOOP_HOME/conf/hadoop-env.sh 文件，添加以下命令行：

```
export HADOOP_OPTS="-agentlib:jdwp=transport=dt_socket,server=y, suspend=y, address=7272"
```

6. 运行 CountersExample.jar 文件，配置本地 weblog_entries_bad_records.txt 文件路径作为输入，并设置一个本地输出路径：

```
$HADOOP_HOME/bin/hadoop jar ./CountersExample.jar com.packt.hadoop.solutions.Counterexample /local/path/to/weblog_entries_bad_ records.txt /local/path/to/weblog_entries_clean.txt
```

你将会看到以下输出：

```
Listening for transport dt_socket at address: 7272
```

7. 在 Eclipse 中打开 **Counters** 工程，并设置远程调试配置。

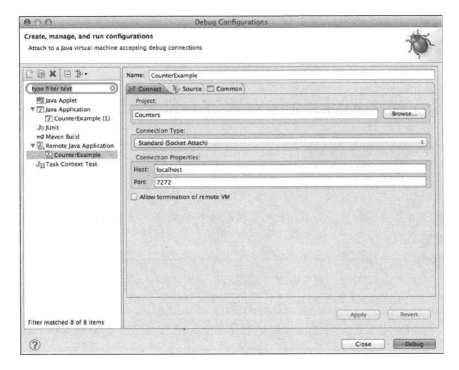

8. 设置新的断点并调试。

工作原理

配置为本地化运行的 MapReduce 作业，是运行在同一个 JVM 实例中。不同于伪分布式模式，该模式下可以调用远程调试器对作业进行调整。配置 `mapred.job.tracker` 值为 `local` 将告知 Hadoop 框架作业将运行在本地模式下。本地模式运行时 `LocalJobRunner` 类发挥作用，负责将 MapReduce 的框架在一个本地化进程中实现。这使本地模式运行和分布式集群上运行作业尽可能一致。使用 `LocalJobRunner` 的一个缺点就是繁重的初始化工作。这意味着，即使启动一个最小的作业也需要花几秒钟来运行。设置 `fs.default.name` 属性值为 `file:///` 来配置作业默认使用本地文件系统来寻找输入/输出文件。在 `hadoop-env.sh` 中添加 `export HADOOP_OPTS="-agentlib:jdwp=transport=dt_socket, server=y,suspend=y, address = 7272"` 是设置 JVM 启动时远程监听的端口 `7272`，配置调试及暂停处理。

更多参考

Apache Pig 也提供了用于开发和测试的本地模式。它使用和本地模式运行 MapReduce 作业相同的 `LocalJobRunner` 类。可以通过以下命令启动 Pig 本地模式作业：

```
Pig -x local
```

> **延伸阅读**
>
> ❑ 使用 MRUnit 开发和测试 MapReduce（8.3 节）。

8.5 运行 MapReduce 作业跳过异常记录

使用 Hadoop 来处理大数据时，即使再健壮的程序也迟早会遇到异常数据。如果异常数据处理不当，很容易造成作业的失败。默认情况下 Hadoop 并不会跳过异常数据，但是对于某些应用程序是可以接受不处理小部分数据的。Hadoop 也提供了这样的方法。即使在不允许跳过任何数据的情况下，Hadoop 的跳跃机制也可以用来找出异常数据并记入日志。

> **操作步骤**

1. 在 Map 中设置允许跳过 100 条异常数据，在 `run()` 方法中添加以下代码：
`SkipBadRecords.setMapperMaxSkipRecords(conf, 100);`

2. 在 Reduce 中设置允许跳过 100 条异常数据，在 `run()` 方法中添加以下代码：
`SkipBadRecords.setReducerMaxSkipGroups(conf, 100);`

> **工作原理**

如果跳过设置生效，那么处理中遇到异常数据就会触发跳过。跳过异常数据的机制是同一个任务在运行失败两次以后通过调用 `SkipBadRecords` 类的静态方法开启的。Hadoop 将对输入的数据执行二分查找，标记出异常的记录。需要记住的是，这是一个代价较高的任务，可能需要进行多次尝试。启用跳过异常记录的设置的作业很有可能增加 map 和 reduce 任务尝试的次数。这个可以通过使用 `JobConf.setMaxMapAttempts()` 和 `JobConf.setMaxReduceAttempts()` 方法来实现。

> **更多参考**

默认情况下，跳过异常数据的处理需要在两次失败尝试后才触发。可以通过使用 `SkipBadRecords` 类的 `setAttemptsToStartSkipping()` 方法来改变默认设置。跳过的记录可以通过 `skipbadrecords` 类的 `setSkipOutputPath()` 方法设置输出文件夹。默认情况下跳过的异常记录将被记录到 `_log/skip/` 文件夹中。这些文件被格式化为 Hadoop 序列化文件。我们可以通过下面的命令读取这些文件：

`hadoop fs -text _log/skip/<filename>`

记录跳过也可以通过修改 MapReduce 作业的配置来控制。下表是由 `http://hadoop.`

apache.org/common/docs/r0.20.2/mapred-default.html 提供的相关表。

属性名称	默认值	描述
mapred.skip.attempts.to.start.skipping	2	跳过模式将在第几次任务尝试失败以后启动。当跳过模式启动，任务报告将记下 TaskTraker 需要处理的记录范围。这样失败的 TaskTracker 就知道哪里可能存在异常记录。这样在下一次处理就知道哪些应该被跳过
mapred.skip.map.auto.incr.proc.count	true	若该标识设置为 true，MapRunner 在调用 map 函数后将自增计数器 SkipBadRecords.COUNTER_MAP_PROCESSED_RECORDS。当应用程序处理的记录是异步或者是缓冲输入，这个值必须设置为 false。例如，流计算。在这种情况下应用程序需要自己实现这样的计数器
mapred.skip.reduce.auto.incr.proc.count	true	若该标识设置为 true，框架在调用 reduce 函数后将自增计数器 SkipBadRecords.COUNTER_REDUCE_PROCESSED_RECORDS。当应用程序处理的记录是异步或者是缓冲输入，这个值必须设置为 false。例如，流计算。在这种情况应用程序需要自己实现这样的计数器
mapred.skip.out.dir		如果该值没有设置，跳过的记录默认会写入 _logs/skip 目录下。用户可以通过设置 none 值不记录跳过的记录。
mapred.skip.map.max.skip.records	0	该值是指 mapper 中遇到异常记录，可接受跳过该异常记录附近记录最大数值。当然这个数值包含异常记录本身。若要关闭检测跳过异常记录的特性，可以将该值设置为 0。该框架尝试缩小重试范围直到达到阈值或者所有的任务尝试都跑完。如果将该值设置为 Long.MAX_VALUE 则不需要尝试缩小。不管多少记录（取决于应用程序）被跳过都是允许的

属性名称	默认值	描述
`mapred.skip.map.max.skip.records`	0	该值是指 reduce 中遇到异常记录，可接受跳过该异常记录附近记录最大数值。当然这个数值包含异常记录本身。若要关闭检测跳过异常记录的特性，可以将该值设置为 0。该框架尝试缩小重试范围直到达到阈值或者所有的任务尝试都跑完。如果将该值设置为 `Long.MAX_VALUE` 则不需要尝试缩小。不管多少记录（取决于应用程序）被跳过都是允许的

8.6 在流计算作业中使用 Counters

Hadoop 并不只限于运行由 Java 或者其他 JVM 语言编写的 MapReduce 作业，它还提供了一个通用的流接口。使用该流接口，任何可以读写标准输入输出流的语言都可以用于开发 MapReduce。由于流作业不需要访问 Hadoop 的 Java 类，需要采用不同的方法来访问框架的特性。Hadoop 提供了一个方便且非常有用的功能——Counters。本节将使用一个简单的 Python 程序展示如何给一个流应用添加一个计数器。Python 代码不能直接访问由 Hadoop 框架用于计数的 Java 报告器。取而代之，它将数据写入标准错误的格式中来代表特定的含义。Hadoop 框架将此解读为增加指定计数器的请求。

准备工作

你需要从 Packet 网站 http://www.packtpub.com/support 下载数据集 weblog_entries_bad_records.txt。这个例子将使用本章代码部分提供的 Python 程序 streaming_counters.py。

操作步骤

完成以下步骤，使用 `streaming_counters.py` 程序运行 Hadoop 流作业。

1. 运行以下命令行：

```
hadoop jar $HADOOP_HOME/contrib/hadoop-*streaming*.jar \
-file streaming_counters.py \
-mapper streaming_counters.py \
-reducer NONE \
-input /data/weblogs/weblog_entries_bad_records.txt \
-output /data/weblogs/weblog_entries_filtered.txt
```

调试 | 185

2. 打开浏览器通过 Job Tracker 的 UI 界面查看作业的计数器信息。默认的地址是 `localhost:50030`。向下滚动到 `Completed Jobs` 段,然后定位到 **streaming-counters** 作业。最新的作业排在列表的最底部,一旦定位到作业就可以点击 **Jobid**。

工作原理

Hadoop 框架不断监视符合以下格式的输入/输出:

`reporter:counter:group,counter,value`

如果发现有一个字符串符合上面的格式,Hadoop 框架就会去检测相应的组合计数器是否存在。如果已经存在,则会把当前的值添加给那个值。如果不存在,则会创建相应的组合计数器,同时对其赋值。

Python 代码对 weblog 数据实现了两个有效验证。第一个是验证列数的有效性:

```
if len(cols) < 5:
sys.stderr.write("reporter:counter:BadRecords,\INVALID_NUMBER_OF_COLS,1")
    continue
```

如果一行少于五列,程序就会按照 Hadoop 期望的格式写出标准错误,来实现操作计数器。同样,第二次效验每条记录中的 IP 地址,如果发现有不合法的 IP 地址则添加计数。

```
m = re.match(('^([01]?\\d\\d?|2[0-4]\\d|25[0-5])\\'
                      '.([01]?\\d\\d?|2[0-4]\\d|25[0-5])\\'
                      '.([01]?\\d\\d?|2[0-4]\\d|25[0-5])\\'
                      '.([01]?\\d\\d?|2[0-4]\\d|25[0-5])$'), ip)
if not m:
sys.stderr.write("reporter:counter:BadRecords,INVALID_IP,1")
continue
```

更多参考

流作业可以使用同样的方式设置任务的状态信息。按照下面的格式输出标准信息将会更新任务的状态,并设置为消息。

`reporter:status:message`

延伸阅读

❑ 运行 MapReduce 作业跳过异常记录(8.5 节)。

8.7 更改任务状态显示调试信息

除了计数之外,Hadoop 中的 `Reporter` 类的另一个作用是捕获任务的状态信息。任务的状态信息定期被发送到 Job Tracker。Job Tracker 的用户界面也相应会更新显示当前的状态。默认情况下，任务的状态栏将显示状态。任务的状态可以是下列之一：

- RUNNING（运行）
- SUCCEEDED（成功）
- FAILED（失败）
- UNASSIGNED（未分配）
- KILLED（已杀死）
- COMMIT_PENDING（提交等待）
- FAILED_UNCLEAN（失败未清理）
- KILLED_UNCLEAN（杀死未清理）

在调试一个 MapReduce 作业时，还可以显示一些自定义信息，给出任务运行更多的详细信息。本节将展示如何更新任务的状态。

准备工作

- 下载本章的源代码。
- 导入 StatusMessage 工程。

操作步骤

可以通过作业的上下文类使用 setStatus() 方法来实现更新任务的状态。

```
context.setMessage("user custom message");
```

工作原理

本章的源代码提供一个例子，使用自定义任务状态信息显示每秒处理的行数。

```
public static class StatusMap extends Mapper<LongWritable, Text, LongWritable, Text> {

    private int rowCount = 0;
    private long startTime = 0;

    public void map(LongWritable key, Text value, Context context) throws IOException, InterruptedException{
        //Display rows per second every 100,000 rows
        rowCount++;
        if(startTime == 0 || rowCount % 100000 == 0)
        {
            if(startTime > 0)
```

```
{
                long estimatedTime = System.nanoTime() - startTime;
                context.setStatus("Processing: " + (double)rowCount / ((double)
estimatedTime/1000000000.0) + " rows/second");
                rowCount = 0;
}
                startTime = System.nanoTime();
}
        context.write(key, value);
}
}
```

定义了两个私有类成员变量:`rowCount` 用于保存监测处理的记录行数,`startTime` 用于保存开始处理的时间。map 函数每处理 100 000 行就会更新任务状态——每秒处理的行数。

```
context.setStatus("Processing: " + (double)rowCount / ((double)
estimatedTime/1000000000.0) + " rows/second");
```

每次状态被更新以后,`rowCount` 和 `startTime` 变量值被重置,整个处理又重新开始。该状态被当前处理进程保存在内存中,然后发送到 Task Tracker。等到下一次 Task Tracker 与 Job Tracker 发送心跳时,再将状态发送到 Job Tracker。Job Tracker 一接收到状态消息就将这些信息显示在用户界面上。

8.8 使用 illustrate 调试 Pig 作业

对于计算数 GB 乃至上 TB 数据,涉及连接、过滤、聚合的复杂分布式作业而言,产生好的测试数据是程序开发中最难的部分,或者至少是最乏味的。Apache Pig 提供了一个令人难以置信的强大工具——illustrate,它根据提供的所有输入数据,找出不同的数据流路径。下面的章节将展示使用 `illustrate` 命令的例子。

准备工作

需要安装 Apache Pig 0.10 版本或者更新的版本。你可以从 `http://pig.apache.org/releases.html` 下载。

操作步骤

下面的 Pig 代码将展示一个记录中包含错误的 IP 地址的例子:

```
weblogs = load '/data/weblogs/weblog_entries_bad_records.txt'
    as (md5:chararray, url:chararray, date:chararray, time:chararray,
ip:chararray);

ip_addresses = foreach weblogs generate ip;

bad = filter ip_addresses by not
```

```
(ip matches '^([01]?\\d\\d?|2[0-4]\\d|25[0-5])\\.([01]?\\d\\d?|2[0-4]\\
d|25[0-5])\\.([01]?\\d\\d?|2[0-4]\\d|25[0-5])\\.([01]?\\d\\d?|2[0-4]\\
d|25[0-5])$');

illustrate bad;
```

输出结果显示如下:

```
| weblogs        | md5:chararray                    | url:chararray          | date:chararray | time:
chararray       | ip:chararray                     |
---------------------------------------------------------------------------------------------------
|                | 8372babe9ebf72b32719feca7        | /ythen.html            | 2012-05-10     | 21:40
:20              | 65.392.                          |
|                | baa2f0917c90342e4d7771dbfae5d9   | /fefumrfgkhqlisoke.html| 2012-05-10     | 21:24
:30              | 220.22.74.176                    |

| ip_addresses   | ip:chararray     |
                 | 65.392.          |
                 | 220.22.74.176    |

| bad            | ip:chararray     |
                 | 65.392.          |
```

工作原理

在前面的示例中,对数据进行了无效 IP 地址的过滤。其中无效的 IP 地址的记录数只占到总数的小部分。如果按照传统的抽样方法来创建测试数据,那么采集到的数据将不会包含任何无效的 IP 地址的记录。

Illustrate 的算法在一个 Pig 脚本之上创建了 4 个完整的通道用以生成数据。第一个通道对每个输入部分进行数据抽样并输出用于执行脚本;第二个通道找出并除去执行脚本中具有相同路径的记录;第三通道确认任何可能的路径是否在第一次抽样数据中存在遗漏,如果存在这样的路径,那么 illustrate 就会制造一些假数据模拟余下的路径;第四个通道与第二个类似,除去由第三个通道创建的任何冗余数据。

更多参考

❑ 要了解如何为数据流程序生成测试数据,可访问 http://i.stanford.edu/~olston/publications/sigmod09.pdf。

第 9 章

系统管理

本章我们将介绍:
- 在伪分布模式下启动 Hadoop
- 在分布式模式下启动 Hadoop
- 添加一个新节点
- 节点安全退役
- NameNode 故障恢复
- 使用 Ganglia 监控集群
- MapReduce 作业参数调优

9.1 介绍

本章我们将讨论如何搭建、监控、调优 Hadoop 集群和 MapReduce 作业。我们将回顾 Hadoop 操作的多种模式,描述如何解决 Hadoop 集群出现的问题,并回顾一些比较重要的作业调优参数。

9.2 在伪分布模式下启动 Hadoop

Hadoop 支持以下三种不同的操作模式。

- 单机模式:在这个模式下,Hadoop 只运行在一个节点的一个进程中。

❑ 伪分布式模式:在这个模式下,Hadoop 的服务分布在同一个节点的不同进程中。

❑ 全分布式模式:在这个模式下,Hadoop 的服务分布在多个节点的不同进程中。

本节将介绍如何安装和配置使得 Hadoop 运行在伪分布式模式下。在伪分布式模式下,所有 HDFS 和 MapReduce 的进程都将在同一台主机上运行。伪分布式模式是一个很好的测试环境,可以用来测试你对 HDFS 的操作以及 MapReduce 程序。

准备工作

确认已经安装 Java 1.6、ssh 和 sshd。此外,shh 的守护进程(sshd)应在节点上运行。你可以通过以下的命令验证这些安装的正确性:

```
$ java -version
java version "1.6.0_31"
Java(TM) SE Runtime Environment (build 1.6.0_31-b04)
Java HotSpot(TM) 64-Bit Server VM (build 20.6-b01, mixed mode)

$ ssh
usage: ssh [-1246AaCfgkMNnqsTtVvXxY] [-b bind_address] [-c cipher_spec]
           [-D [bind_address:]port] [-e escape_char] [-F configfile]
           [-i identity_file] [-L [bind_address:]port:host:hostport]
           [-l login_name] [-m mac_spec] [-O ctl_cmd] [-o option] [-p port]
           [-R [bind_address:]port:host:hostport] [-S ctl_path]
           [-w tunnel:tunnel] [user@]hostname [command]

$ service sshd status
openssh-daemon (pid 2004) is running...
```

操作步骤

执行以下步骤来启动 Hadoop 伪分布式模式。

1. 创建一个 Hadoop 用户。Hadoop 伪分布式模式并不能解决什么特别的需求,但是它是一种常用和良好的安全实践。确认 `Java_HOME` 环境变量已经设置了系统 Java 安装的目录:

```
# useradd hadoop
# passwd hadoop
# su - hadoop
$ echo $JAVA_HOME
$ /usr/java/jdk1.6.0_31
```

2. 生成 ssh 公钥和私钥,配置 Hadoop 账户可以实现免密码登录到本机。当提示需要输入密码的时候,直接输入回车,确保没有密码被使用:

```
$ su - hadoop
$ ssh-keygen -t rsa
```

3. 添加公钥到 authorized key 文件列表中：

 如果有多个节点，则需要复制到集群的每个节点上。

```
$ ssh-copy-id -i /home/hadoop/.ssh/id_rsa.pub hadoop@localhost
```

4. 测试免密码登录。不用输入密码你就可以通过 `ssh localhost` 登录到本机：

```
$ ssh localhost
```

5. 从 Hadoop 官网 http://hadoop.apache.org 下载一个安装包，本文使用的是 Hadoop 0.20.x 版本。使用 Hadoop 用户账户进行安装：

```
$ su -hadoop
$ tar -zxvf hadoop-0.20.x.tar.gz
```

6. 修改 Hadoop 包解压后 `conf` 目录下的配置文件。修改这些配置允许 Hadoop 在伪分布模式下运行：

```
$ vi conf/core-site.xml
<configuration>
  <property>
    <name>fs.default.name</name>
    <value>hdfs://localhost:8020</value>
  </property>
</configuration>
$ vi conf/hdfs-site.xml
<configuration>
  <property>
    <name>dfs.replication</name>
    <value>1</value>
  </property>
</configuration>
$ vi conf/mapred-site.xml
<configuration>
  <property>
    <name>mapred.job.tracker</name>
    <value>localhost:8021</value>
  </property>
</configuration>
```

7. 格式化 Hadoop NameNode：

```
$ bin/hadoop namenode -format
```

8. 启动 Hadoop HDFS 和 MapReduce 服务：

```
$ bin/start-all.sh
```

9. 验证服务是否成功启动，通过访问 http://localhost:50070 和 http://localhost:50030 分别查看 NameNode 状态页面和 JobTracker 页面是否正常。你可以通过执行

`bin/stop-all.sh` 脚本停止 Hadoop 的服务。

工作原理

步骤 1 至步骤 4 为在一个单独的节点上使用 ssh 配置免密码登录。

接着我们下载了 Hadoop 的一个安装包，并且通过修改配置使其运行在伪分布模式下。`fs.default.name` 属性采用 URI 地址格式，它将告诉 Hadoop，HDFS 文件系统的服务地址。在本例子中，HDFS 将在本机的 8020 端口启动。接着我们通过修改 `dfs.replication` 的属性值将 HDFS 的备份因子设置为 1。因为我们在一个节点上运行了 Hadoop 所有的服务，没必要进行多备份。如果我们设置了，那么所有的复制信息也将存在这个单点上。最后我们将 `mapred.job.Tracker` 设置为 `localhost:8021`。该属性将决定 JobTracker 服务的地址。

最后我们格式化 NameNode 并启动 Hadoop 服务。在配置完一个新的集群后你需要对 NameNode 进行格式化。格式化完 NameNode，Hadoop 集群的数据将被清空。

更多参考

默认情况下 Hadoop 是运行在单机模式下。在单机模式下不需要启动任何 Hadoop 服务。另外输入和输出文件都是在本地文件系统而不是 HDFS。若要在本地模式下运行 MapReduce，可以使用初始的配置信息。在本地文件系统创建输入文件夹并且使用 Hadoop shell 脚本：

```
$ mkdir input
$ cp somefiles*.txt input/
$ path/to/hadoop/bin/hadoop jar myjar.jar input/*.txt output
```

延伸阅读

- 在分布式模式下启动 Hadoop（9.3 节）。

9.3 在分布式模式下启动 Hadoop

在前一节中我们已经介绍过了，Hadoop 支持以单机模式、伪分布式模式、全分布式模式三种不同的模式进行操作。

本章将描述如何配置 Hadoop 运行在全分布式模式下。在全分布式模式下，HDFS 和 MapReduce 服务将运行在不同的机器。一种经典的安装方式，是将 NameNode 和 JobTracker 服务运行在一个专用的节点上，Secondary NameNode 服务运行在另一个专用的机器上。集群中剩余的节点都运行 DataNode 和 TaskTracker 服务。

准备工作

本节假定上节步骤 1 至步骤 5 都已经配置完毕。集群中的每个节点上都需创建一个 `hadoop` 账户。此外，在第 2 步中生成的 `rsa` 公共键必须分发并安装到集群的所有节点中，可以通过 `ssh-copy-id` 命令完成。最后 Hadoop 安装包需要在集群的每个节点上进行解压和部署。

我们现在讨论一下分布式模式下的参数配置。假定你的集群使用了下面的配置。

服务器名称	描述	机器数量
head	运行 NameNode、JobTracker 服务	1
secondary	运行 Seconday NameNode 服务	1
worker(n)	运行 TaskTracker 和 DataNode 服务	3 个或者更多

操作步骤

执行以下几个步骤，在分布式模式下启动 Hadoop。

1. 在所有节点上更改以下配置文件：

```
$ vi conf/core-site.xml
<configuration>
  <property>
    <name>fs.default.name</name>
    <value>hdfs://head:8020</value>
  </property>
</configuration>

$ vi conf/hdfs-site.xml
<configuration>
  <property>
    <name>dfs.replication</name>
    <value>3</value>
  </property>
</configuration>

$ vi conf/mapred-site.xml
<configuration>
  <property>
    <name>mapred.job.tracker</name>
    <value>head:8021</value>
  </property>
</configuration>
```

2. 修改 head 节点上的 `masters` 和 `slaves` 配置文件。`masters` 配置文件包含了 Secondary NameNode 服务主机名。`slaves` 文件包含了所有运行了 TaskTraker 和 datanode 服务的主机名：

```
$ vi conf/masters
secondary
$ vi conf/slaves
worker1
worker2
worker3
```

3. 在 head 节点上执行 NameNode 格式化：

```
$ bin/hadoop namenode -format
```

4. 在 head 节点上使用 hadoop 账户启动 hadoop 服务：

```
$ bin/start-all.sh
```

5. 确认各个节点上运行的服务。

- head 节点：NameNode 和 JobTracker 服务启动。
- secondary 节点：Secondary NodeNode 服务启动。
- worker 节点：DataNode 和 TaskTracker 服务启动。

工作原理

首先我们修改了集群上每个节点上的 core-site.xml、hdfs-site.xml 和 mapred-site.xml。这些配置修改将决定每个节点上可以启动的服务，以及 NameNode 和 JobTracker 服务的地址。此外我们将 HDFS 的备份因子调整为 3。因为我们拥有 3 个以上的可用节点，我们将备份因子从 1 调整为 3，以保证数据的高可用性，当某个 worker 节点发生错误数据乃可用。

更多参考

Secondary NameNode 并不是一定要安装在一个独立的节点上。你可以将 Seconday NameNode 和 NameNode、JobTracker 服务启动在同一台机器上。如果打算这么做，那么停止集群，在主节点上修改 master 配置，重启所有的服务：

```
$ bin/stop-all.sh
$ vi masters
head
$ bin/start-all.sh
```

另外一个参数配置在你的集群规模增长的时候会派上用场,或者当你需要对集群进行维护时，可以在 mapred-site.xml 配置文件中添加排除列表参数。通过在 mapred-site.xml 添加下面列表，你可以列出不准连接到 Name Node（dfs.hosts.exclude）和/或 Job Tracker（mapred.hosts.exclude）的节点列表。之后，当我们谈到从集群中解除一个节点时，还会用到这些配置参数。

```
<property>
    <name>dfs.hosts.exclude</name>
    <value>/path/to/hadoop/dfs_excludes</value>
    <final>true</final>
</property>
<property>
    <name>mapred.hosts.exclude</name>
    <value>/path/to/hadoop/mapred_excludes </value>
    <final>true</final>
</property>
```

创建两个空文件，分别命名为 `dfs_excludes` 和 `mapred_excludes`：

```
$ touch /path/to/hadoop/dfs_excludes
$ touch /path/to/hadoop/mapred_excludes
```

启动集群：

```
$ bin/start-all.sh
```

延伸阅读

- 添加一个新节点（9.4 节）。
- 节点安全退役（9.5 节）。

9.4 添加一个新节点

Hadoop 支持在不停止或重启集群的情况下添加新节点。本节将介绍在线上的集群中添加新节点的步骤。

准备工作

确定你已经配置好并启动了一个 Hadoop 集群。此外确定你已经按照 9.3 节介绍的更新配置文件。

使用下面的词条作为集群的描述。

服务器名称	描述	机器数量
head	运行 NameNode、JobTracker 服务	1
secondary	运行 Seconday NameNode 服务	1
worker(n)	运行 TaskTracker 和 DataNode 服务	3 个或者更多

操作步骤

完成以下步骤，为已存在的集群添加一个新节点。

1. 在 head 节点上，修改 `slaves` 配置文件添加新节点的主机域名：

```
$ vi conf/slaves
worker1
worker2
worker3
worker4
```

2. 登录到新节点并启动 DataNode 和 TaskTracker 服务：

```
$ ssh hadoop@worker4
$ cd /path/to/hadoop
$ bin/hadoop-daemon.sh start datanode
$ bin/hadoop-daemon.sh start tasktracker
```

工作原理

我们在 head 节点上更新了 `slaves` 配置文件，告诉 Hadoop 框架集群中存在一个新的节点。但是，这个文件只是在服务启动的时候才被读取（如通过执行 `bin/start-all.sh` 脚本）。为了在不重启 Hadoop 服务的前提下添加一个新节点，我们需要登录到新节点手动启动 DataNode 和 TaskTracker 服务。

在集群下次重启的时候会自动启动 DataNode 和 TaskTracker 服务。

更多参考

当你给集群添加一个新的节点，整个集群存储负载可能就不平衡。HDFS 不会自动地将其他节点的数据重新分布到新节点上保证数据存储的平衡。我们可以在 head 节点上执行以下命令平衡集群中的数据存储。

```
#bin/start-balancer.sh
```

平衡 Hadoop 是一个网络 IO 密集型作业。想象一下我们可能需要移动大约 1TB 的数据，当然取决新增加节点的数量。在平衡的过程中，作业运行的效率会有所下降，因此应当定期对集群进行平衡。

延伸阅读

❏ 节点安全退役（9.5 节）。

9.5 节点安全退役

将集群中的节点进行摘除是十分常见的操作。硬件可能出错或者机器可能需要升级等等。在本节我们将展示如何安全地将 Hadoop 集群中一个工作的节点摘除。

准备工作

假设你的 Hadoop 集群已经配置并且运行着。打开 `mapred-site.xml`，你会看到如下配置。

```
<property>
    <name>dfs.hosts.exclude</name>
    <value>/path/to/hadoop/dfs_excludes</value>
    <final>true</final>
</property>
<property>
    <name>mapred.hosts.exclude</name>
    <value>/path/to/hadoop/mapred_excludes </value>
    <final>true</final>
</property>
```

此外，在 head 节点的 Hadoop 安装目录下存在两个文件：`dfs_excludes` 和 `mapred_excludes`。

操作步骤

执行以下步骤，实现将 Hadoop 集群中的一个节点进行退役。

1. 修改 head 节点的 `dfs_excludes` 和 `mapred_excludes` 文件，添加需要摘除的节点主机名称：

```
$ vi /path/to/hadoop/dfs_excludes
worker1
$ vi /path/to/hadoop/mapred_excludes
worker1
```

2. 通知 NameNode 重新读取排除列表，并断开工作节点使其退役：

```
$ hadoop dfsadmin -refreshNodes
```

3. 通知 JobTracker 重新读取排除列表，并断开工作节点使其退役：

```
$ hadoop mradmin -refreshNodes
```

4. 检查退役操作的状态：

```
$ hadoop dfsadmin -report
```

工作原理

首先，我们将需要退役的节点主机名称添加到 `dfs_excludes` 和 `mapred_excludes` 文件中。接着我们执行 `Hadoop dfsadmin-refreshnodes` 命令通知 NameNode 断开与所有在 `dfs_excludes` 文件中列出的主机服务器之间的连接。同样的，我们执行 `Hadoop mradmin-refreshnodes` 命令通知 JobTracker 停止使用在 `mapred_excludes` 文件中列出的节点的 TaskTracker 服务。

9.6 NameNode 故障恢复

NameNode 是 Hadoop 中最重要的服务。它保存着集群中所有数据块的存储位置，也保存着分布式文件系统的状态。当然 NameNode 发生故障，可以通过 Secondary NameNode 产生的上一个检测点进行恢复。有一点需要明确知道，Secondary NameNode 并不是 NameNode 的备份。它定期执行检测点操作，所以从 Secondary NameNode 那边恢复过来的数据不是最新的。但是，使用旧的文件系统信息总比没恢复好。

准备工作

假定运行 NameNode 的那台主机服务出现故障，并且 Secondary NameNode 运行在另外一台独立的主机上。此外，在 `core-default.xml` 配置文件中已经配置了 `fs.checkpoint.dir` 属性。这个配置告诉 Secondary NameNode 需要将检测点保存到本地文件系统的哪个路径。

操作步骤

执行以下步骤实现 NameNode 的故障恢复。

1. 停止 Secondary NameNode 服务：

```
$ cd /path/to/hadoop
$ bin/hadoop-daemon.sh stop secondarynamenode
```

2. 配置一台新机器作为新的 NameNode。这台机器一定要安装好 Hadoop，并且配置和之前的 NameNode 必须一致同时还需要配置 ssh 免密码登录。此外，这台机器的 IP 和主机名称必须和原先的 NameNode 一致。

3. 复制 Secondary NameNode 上 `fs.checkpoint.dir` 下的内容到新 NameNode 的 `dfs.name.dir` 目录下。

4. 在新机器上启动 NameNode 服务：

```
$ bin/hdoop-daemon.sh start namenode
```

5. 在 Secondary NameNode 上启动 Secondary NameNode 服务：

```
$ bin/hdoop-daemon.sh start secondarynamenode
```

6. 通过访问 NameNode 信息页面 `http://head:50070` 来确认 NameNode 服务已成功启动。

工作原理

我们首先登入 Secondary NameNode 服务器，并停止 Secondary NameNode 服务。接着我们创建一台与旧 NameNode 一模一样的机器。再将 checkpoint 和 edit 文件从 Secondary NameNode 复制到新 NameNode 上。这将允许我们恢复文件系统的状态、元数据和操作到上个检测点的时候。最后我们启动新的 NameNode 和 Secondary NameNode。

更多参考

在某些场合下使用旧的数据是不可接受的。相应的存在另外一种解决方案，构建第三方离线存储，NameNode 可以将 image 和 edit 文件写入。在这种方式下，当 NameNode 机器硬件存在错误，你可以恢复最新的文件系统而不需要依靠 Secondary NameNode 的快照。

第一步，指定一台新的机器保存 NameNode 的 image 和 edit 文件备份。接着在 NameNode 上挂载备份机。最后，在运行 NameNode 服务的那台机器上修改 `hdfs-site.xml` 使其写入本地文件和挂载备份机上：

```
$ cd /path/to/hadoop
$ vi conf/hdfs-site.xml
<property>
    <name>dfs.name.dir</name>
    <value>/path/to/hadoop/cache/hadoop/dfs, /path/to/backup</value>
</property>
```

现在 NameNode 就会把文件系统的元信息既写入 `/path/to/hadoop/cache/hadoop/dfs` 也写入挂载盘 `/path/to/backup`。

9.7 使用 Ganglia 监控集群

Ganglia 是一个监控系统，设计用于对集群和网格进行监控。Hadoop 可以配置周期性发送指标（metric）信息到 Ganglia 的守护进程，这对我们监控和诊断 Hadoop 集群的健康是很有帮助的。本节我们将会解析如何配置 Hadoop，使其发送指标信息给 Ganglia 监控进程。

准备工作

确保你在 Hadoop 集群的所有节点上已经安装了 Ganglia 3.1 版本或者其他更新的版本。

在所有的工作节点上都必须启动 Ganglia 监控守护进程（gmond）。你必须至少在一个节点上运行 Ganglia 元数据服务（gmetad），并且在另一个节点上运行 Ganglia 的前端网页服务。

以下是一个已经修改过的 gmond.conf 文件，可以用于 gmond 守护进程：

```
cluster {
  name = "Hadoop Cluster"
  owner = "unspecified"
  latlong = "unspecified"
  url = "unspecified"
}

host {
  location = "my datacenter"
}
udp_send_channel {
  host = mynode.company.com
  port = 8649
  ttl = 1
}
udp_recv_channel {
  port = 8649
}

tcp_accept_channel {
  port = 8649
}
```

此外，还需要确认 Ganglia 元信息守护进程配置文件包含了你的集群并设置了数据源。例如，修改 gmeta.conf 文件添加 Hadoop 集群为一个数据源：

```
data_source "Hadoop Cluster" mynode1.company.com:8649 mynode2.company.com:8649 mynode3.company.com:8649
```

操作步骤

执行下面的步骤，完成使用 Ganglia 监控集群信息。

1. 修改 Hadoop 配置文件中的 hadoop-metics.properties 文件。若该文件不存在则创建一个同名的新文件。每个节点下的配置文件都要修改：

```
$ vi /path/to/hadoop/hadoop-metrics.properties
dfs.class=org.apache.hadoop.metrics.ganglia.GangliaContext31
dfs.period=10
dfs.servers=mynode1.company.com:8649

mapred.class=org.apache.hadoop.metrics.ganglia.GangliaContext31
mapred.period=10
mapred.servers=mynode1.company.com 8649

jvm.class=org.apache.hadoop.metrics.ganglia.GangliaContext31
```

```
jvm.period=10
jvm.servers=mynode1.company.com:8649

rpc.class=org.apache.hadoop.metrics.ganglia.GangliaContext31
rpc.period=10
rpc.servers=mynode1.company.com 8649
```

2. 重启 Ganlia 元守护进程服务。

3. 重启 Hadoop 集群：

```
$ cd /path/to/hadoop
$ bin/stop-all.sh
$ bin/start-all.sh
```

4. 通过 Ganglia 前端 Web 验证 Ganglia 是否成功收集 Hadoop 指标信息。

工作原理

`gmond` 主要负责从各个节点上收集指标信息。接着，这些收集到的指标信息会被汇总到 `gmetad` 上。最后 Ganglia Web 界面会请求 `gmetad` 上聚合的指标信息，通过 XML 格式返回数据并显示到界面上。

9.8 MapReduce 作业参数调优

Hadoop 框架是非常灵活的，可以使用大量的配置参数进行调优。在本节中，我们将讨论那些可以在 MapReduce 作业中设置的不同的配置参数的功能以及使用目的。

准备工作

确定你的 MapReduce 作业继承了 Hadoop `Configuration` 类和实现了 Hadoop `Tool` 的接口，就像本书中写过的任何 MapReduce 应用一样。

操作步骤

完成下面步骤，调节 MapReduce 作业的参数。

1. 确保你的 MapReduce 作业类继承 Hadoop `Configuration` 类并且实现了 `Tool` 接口。

2. 使用 `ToolRunner.run()` 这个静态方法运行你的 MapReduce 作业，如下面的例子所示：

```
public static void main(String[] args) throws Exception {
    int exitCode = ToolRunner.run(new MyMapReduceJob(), args);
    System.exit(exitCode);
}
```

3. 检查下表中关于 Hadoop 作业的属性以及参考值:

属性名称	参考值	描述
`mapred.reduce.tasks`	整数（0~N）	设置需要启动的 reducer 个数
`mapred.child.java.opts`	JVM 键值对	这些参数是设置给每个任务的 JVM。比如设置堆大小为 1 GB，你需要设置该值为`'-Xmx1GB'`
`mapred.map.child.java.opts`	JVM 键值对	这些参数是设置给 map 任务的 JVM
`mapred.reduce.child.java.opts`	JVM 键值对	这些参数是设置给 reduce 任务的 JVM
`mapred.map.tasks.speculative.execution`	布尔型（true 或 false）	是否在 map 阶段启动预测机制，`true` 为启动，如果某个任务与其他任务比较运行得不够好，则会在不同的节点上再执行相同的 map 任务，该属性在第 1 章已经讨论过
`mapred.reduce.tasks.speculative.execution`	布尔型（true 或 false）	是否在 reduce 阶段启动预测机制，`true` 为启动，如果某个任务与其他任务比较运行的不够好，则会在不同的节点上再执行相同的 reduce 任务
`mapred.job.reuse.jvm.num.tasks`	整数（-1,1-N）	任务 JVM 重用的数量。设置为 1 表示每个每个任务使用 1 个 JVM。设置为-1 表示 1 个 JVM 可以允许无限的任务。设置该参数可以调优小任务，重用 JVM 避免每个任务都重新启动一个 JVM
`mapred.compress.map.output`	布尔型（true 或 false）	这三个参数用于压缩 map 任务的输出结果
`mapred.output.compression.type`	字符串（NONE、RECORD 或 BLOCK）	
`mapred.map.output.compression.codec`	字符串（压缩算法类名）	

续表

属性名称	参考值	描述
`mapred.output.compress`	布尔型（true 或 false）	
`mapred.output.compression.type`	字符串（NONE、RECORD 或者 BLOCK）	这三个参数用于压缩 MapReduce 作业的输出结果
`mapred.output.compression.codec`	字符串（压缩算法类名）	

4. 运行 MapReduce 作业使用自定义属性，例如运行一个作业配置 5 个 reducer：

```
$ cd /path/to/hadoop
$ bin/hadoop -jar MyJar.jar com.packt.MyJobClass -Dmapred.reduce.tasks=5
```

工作原理

当一个作业类继承 Hadoop `Configuration` 类并且实现了 Hadoop `Tool` 接口，`ToolRunner` 类会自动处理以下 Hadoop 参数：

参数/标志	目的
`-conf`	指定参数配置文件的路径
`-D`	用于指定将被加载到作业配置中键值属性
`-fs`	用于指定 NameNode 的主机端口
`-jt`	用于指定 JobTracker 的主机端口

在本节的例子中，`ToolRunner` 会自动将 `-D` 后面指定的所有参数加载到 Hadoop 作业的 XML 配置文件中。

第 10 章

使用 Apache Accumulo 进行持久化

本章我们将介绍：
- 在 Accumulo 中设计行键存储地理事件
- 使用 MapReduce 批量导入地理事件数据到 Accumulo
- 设置自定义字段约束 Accumulo 中的地理事件数据
- 使用正则过滤器限制查询结果
- 使用 SumCombiner 计算同一个键的不同版本的死亡数总和
- 使用 Accumulo 实行单元级安全的扫描
- 使用 MapReduce 聚集 Accumulo 中的消息源

10.1 介绍

大数据存储是一个日益热门的话题。软件工程在面对数据频繁扩展的需求时不得不花高价购买昂贵的关系数据库商业许可证，或者更糟，不得不依赖那些水平扩展性能差的解决方案。在过去的几年中，我们已经看到了许多可行的开源数据库解决方案用来帮助管理大量的结构化和非结构化数据。Apache Accumulo 的灵感来自谷歌 **BigTable** 的设计方法，并支持水平扩展，支持 Apache Hadoop 的分布式列数据的持久化。BigTable 设计的细节可以参考 http://research.google.com/archive/bigtable.html。本章将用几节来介绍常见的数据库查询和加载任务，并展示 Accumulo 帮助简化实现的许多特性。

10.2 在 Accumulo 中设计行键存储地理事件

武装冲突位置事件数据（ACLED）是对一个广泛的地理区域内发生的独立事件进行收集。本章将展示我们如何利用 Accumulo 的关键词排序将事件按照地理位置范围进行分组。此外，每个地理位置范围将按照事件发生的顺序来存放。具体来说，本节中的代码展示了如何从我们的记录中创建 ACLED 键值的逻辑。为了验证生成的键值符合我们的预期，我们会将代码编译，并使用一些样本数据进行单元测试。

准备工作

为了进行单元测试，你需要在系统环境的 classpath 中配置 TestNG。熟悉 TestNG 测试的一些基本接口，对进行单元测试很有帮助。

本节将使用特定的四叉树数据结构，将地理空间数据转化为索引范围。如果对 Z 曲线（又名莫尔顿曲线）有一定了解，将有助于我们使用二维空间数据来构建这样的四叉树结构。

操作步骤

按照下面的步骤实现一个按照地理位置和时间倒序的键值生成器。

1. 打开你的 Java IDE 编辑器。

2. 按照下面的内容创建包 example.accumulo 和接口 RowIDGenerator.java：

```
package examples.accumulo;

import javax.security.auth.login.Configuration;
import java.io.IOException;

public interface RowIDGenerator {
public String getRowID(String[] parameters)
        throws IllegalArgumentException;
}
```

3. 同样在 example.accumulo 包下创建类 ACLEDRowIDGenerator.java：

```
package examples.accumulo;

import java.text.DateFormat;
import java.text.ParseException;
import java.text.SimpleDateFormat;
import java.util.Date;
import java.util.regex.Pattern;

public class ACLEDRowIDGenerator implements RowIDGenerator {
```

```
private DateFormat dateFormat = new
                           SimpleDateFormat("yyyy-MM-dd");
private static final Pattern decimalPattern = Pattern.compile("[.]");
```

4. 编写 `getRowID()` 方法处理传入的 `String[]` 参数。

```
@Override
public String getRowID(String[] parameters)
        throws IllegalArgumentException {
    if(parameters.length != 3)
        throw new IllegalArgumentException("Required:{lat, lon, dtg}")
    StringBuilder builder = new StringBuilder();
    builder.append(getZOrderedCurve(parameters[0],
                                    parameters[1]));
    builder.append("_");
    builder.append(getReverseTime(parameters[2]));
    return builder.toString();
}
```

5. 创建 `getZOrderedCurve()` 用于构建 rowID 的地理位置部分。将方法设置为公有的是为了方便进行单元测试。

```
public String getZOrderedCurve(String lat, String lon)
        throws IllegalArgumentException {
    StringBuilder builder = new StringBuilder();
    lat = cleanAndValidatePoint(lat);
    lon = cleanAndValidatePoint(lon);
    int ceiling = Math.max(lat.length(), lon.length());
    for (int i = 0; i < ceiling; i++) {
        if(lat.length() <= i) {
            builder.append("0");
        } else {
            builder.append(lat.charAt(i));
        }
        if(lon.length() <= i) {
            builder.append("0");
        } else {
            builder.append(lon.charAt(i));
        }
    }
    return builder.toString();
}
```

6. 私有方法 `cleanAndValidatePoint` 将帮助验证和审查经纬度转化为一个合适的格式用于 Z 排序:

```
private String cleanAndValidatePoint(String point)
        throws IllegalArgumentException {

    String[] pointPieces = decimalPattern.split(point);
    if(pointPieces.length > 2) {
        throw new IllegalArgumentException("Malformed point: " + point);
    }
```

```java
            String integralStr = null;
            int integral = 0;
            try {
                //offset any negative integral portion
                integral = Integer.parseInt(pointPieces[0]) + 90;
                if(integral > 180 | integral < 0) {
                    throw new IllegalArgumentException("Invalidintegral: " +
    integral + " for point: " + point);
                }
                integralStr = "" + integral;
                if(pointPieces.length > 1)
                    integralStr += Integer.parseInt(pointPieces[1]);
                if(integral < 10)
                    integralStr = "00" + integralStr;
                else if (integral >= 10 && integral < 100)
                    integralStr = "0" + integralStr;
                return integralStr;
            } catch (NumberFormatException e) {
                throw new IllegalArgumentException("Point: " +
                    point + " contains non-numeric characters");
            }
        }
```

7. getReverseTime()公有方法用于创建行键的时间戳，设置为公有主要是为了方便单元测试。

```java
        public long getReverseTime(String dateTime)
                throws IllegalArgumentException {
            Date date = null;
            try {
                date = dateFormat.parse(dateTime);
            } catch (ParseException e) {
                throw new IllegalArgumentException(dateTime +
                            "Could not be parsed to a " +
                        "valid date with the supplied DateFormat " +
    dateFormat.toString());
            }
            return Long.MAX_VALUE - date.getTime();
        }
    }
```

8. 在 examples.accumulo 包下创建一个 TestNG 单元测试类 ValidatingKeyGenTest.java，代码如下：

```java
package examples.accumulo;

import org.apache.hadoop.hbase.thrift.generated.IllegalArgument;
import org.testng.annotations.BeforeClass;
import org.testng.annotations.Test;
import static org.testng.Assert.*;

import java.text.ParseException;
```

```java
import java.text.SimpleDateFormat;
import java.util.Date;

public class ValidatingKeyGenTest {

    private ACLEDRowIDGenerator keyGen;
    private SimpleDateFormat dateFormatter = new
                      SimpleDateFormat("yyyy-MM-dd");
```

9. 使用`@BeforeClass`注释创建一个`ACLEDRowIDGenerator`的实例。

```java
    @BeforeClass
    public void setup() {
        keyGen = new ACLEDRowIDGenerator();
    }
```

10. 添加单元测试方法`validZOrder()`。

```java
    @Test
    public void validZOrder() {
        try {
            // +90 = 123.22,134.55
            String zpoint = keyGen.getZOrderedCurve("33.22", "44.55");
            assertEquals(zpoint, "1123342525");

            // +90 = 123, 134.55
            zpoint = keyGen.getZOrderedCurve("33", "44.55");
            assertEquals(zpoint, "1123340505");

            // +90 = 123.55, 134
            zpoint = keyGen.getZOrderedCurve("33.55", "44");
            assertEquals(zpoint, "1123345050");

            // +90 = 123.1234, 134.56
            zpoint = keyGen.getZOrderedCurve("33.1234","44.56");
            assertEquals(zpoint, "11233415263040");

            // +90 = 00.11, 134.56
            zpoint = keyGen.getZOrderedCurve("-90.11", "44.56");
            assertEquals(zpoint, "0103041516");

            // +90 = 005.11, 134.56
            zpoint = keyGen.getZOrderedCurve("-85.11", "44.56");
            assertEquals(zpoint, "0103541516");

            // +90 = 011.11, 134.56
            zpoint = keyGen.getZOrderedCurve("-79.11", "44.56");
            assertEquals(zpoint, "0113141516");

            // +90 = 095, 134.56
```

```
            zpoint = keyGen.getZOrderedCurve("5", "44.56");
            assertEquals(zpoint, "0193540506");


        } catch (Exception e) {
            fail("EXCEPTION fail: " + e.getMessage());
        }
    }
```

11. 添加单元测试方法 `invalidZOrder()`。

```
    @Test
    public void invalidZOrder() {
        String zpoint = null;
        try {
            zpoint = keyGen.getZOrderedCurve("98.22", "33.44");
            fail("Should not parse. Too big an integralvalue.");
        } catch (IllegalArgumentException e) {
            assertTrue(e.getMessage().contains("invalidintegral"));
        }

        try {
            zpoint = keyGen.getZOrderedCurve("78.22", "-91.44");
            fail("Should not parse. Too big an integralvalue.");
        } catch (IllegalArgumentException e) {
            assertTrue(e.getMessage().contains("invalidintegral"));
        }

        try {
            zpoint = keyGen.getZOrderedCurve("332.22.33","33.44.33.22");
            fail("Should not parse. Too many split values.");
        } catch (IllegalArgumentException e) {
            assertTrue(e.getMessage().contains("Malformedpoint"));
        }

        try {
            zpoint = keyGen.getZOrderedCurve("33.22a", "33.33");
            fail("Should not parse. Contains bad characters.");
        } catch (IllegalArgumentException e) {
            assertTrue(e.getMessage().contains("contains nonnumericcharacters"));
        }

        try {
            zpoint = keyGen.getZOrderedCurve("33.22", "3c.33");
            fail("Should not parse. Contains bad characters.");
        } catch (IllegalArgumentException e) {
            assertTrue(e.getMessage().contains("contains nonnumericcharacters"));
        }
    }
```

12. 添加单元测试方法 `testInvalidReverseTime()`。

```
@Test
public void testValidReverseTime() {
    String dateStr = "2012-05-23";
    long reverse = keyGen.getReverseTime(dateStr);
    try {
        Date date = dateFormatter.parse(dateStr);
        assertEquals(reverse, (Long.MAX_VALUE - date.getTime()));
    } catch (ParseException e) {
        fail(e.getMessage());
    }
}
```

13. 添加单元测试方法 `testInvalidReverseTime()`。

```
@Test
public void testInvalidReverseTime() {
    try {
        long reverse = keyGen.getReverseTime("201a-22-22");
        fail("Should not reverse invalid date for DateFormat");
    } catch (IllegalArgumentException e) {
        assertTrue(e.getMessage().contains("could not be
parsed to a valid date with the supplied DateFormat"));
    }
}
```

14. 添加单元测试方法 `testFullKey()`。

```
@Test
public void testFullKey() {
    try {
        String dateStr = "2012-03-13";
        Date date = dateFormatter.parse(dateStr);
        long reverse = Long.MAX_VALUE - date.getTime();

        // +90 = 123.55, 156.77
        String key = keyGen.getRowID(new String[]{"33.55", "66.77",
dateStr});
        assertEquals(key, "1125365757_" + reverse);
    } catch (ParseException e) {
        fail(e.getMessage());
    } catch (IllegalArgumentException e) {
        fail(e.getMessage());
    }
}
```

15. 在你的开发环境下运行单元测试，所有的单元测试都应该通过。

工作原理

此代码将作为生成地理位置/时间倒排行键的基础，它作为一个独立的模块存在，不与加载数据到 Accumulo 的任何代码存在耦合。它是专门设计来构造可以按照特别顺序保存到 Accumulo 表中的行键的。

  首先，我们定义了一个通用的接口 `RowIDGenerator.java`，它可以在构建不同键生成器时重用。所有的实现类都必须实现一个简单的方法 `getRowId()`。它以任意的字符串数组作为输入，并返回一个字符串表示 rowID。如果发生错误，将会抛出一个 `IllegalArgumentException` 异常。类 `ACLEDRowIDGenerator.java` 需要输入至少三个字符串的数组。然后我们开始构建 Z 曲线结构为 RowID 生成策略做准备。

  `getZOrderedCurve()` 方法以纬度和经度的字符串作为参数。使用纬度/经度点构建高效的四叉树要求使用的点严格遵循格式化规则，因此在我们洗牌之前就需要通过 `cleanAndValidatePoint()` 函数进行验证和格式化。

  `cleanAndValidatePoint()` 方法首先将小数按照小数点进行分隔，左边为整数部分，右边为小数部分。一个点不一定要包含小数部分，但是一定要包含整数部分。此外不应该拥有多个小数部分。因此如果按照小数点进行分割返回不是 1 个或者 2 个元素的数组，我们将会抛出一个异常 `IllegalArgumentException`。接着我们将每个点的值加上 90 的偏移量，避免出现负数，否则就会打乱 Z 曲线的含义。在应用偏移之后如果包含一个大于 180 或者小于 0 的点，我们可以得出结论，该点开始大于 90 或者小于-90 的数。不符合这两个条件都是无效点，我们将抛出一个 `IllegalArgumentException` 异常来指示这种情况。如果点通过验证，那就可以开始格式化并解析 Z 曲线。根据点的长度，我们需要在前面补零，使得整数部分长度是 3。这样我们研究 `getZOrderCurve()` 方法如何使用结果更有意义。如果适用，将点小数部分追加到格式化的字符串后面表示没有小数点的位置。任何时候捕获一个 `NumberFormatException` 异常就会抛出一个 `IllegalArgumentException` 异常。

  一旦纬度和经度都被正确解析，我们就可以开始洗牌建立四叉树。我们将比较所有的经度和纬度，从中取出最大的两个值作为循环控制体的最大值。从 0 到最大开始循环，以纬度值开始，输出斜杠，在斜杠后输出经度值。我们应该保持两个值的长度一致，如果不足则在后面补 0。这将有助于生成一个统一的纬度/经度键值对，无论经度和纬度之间存在多大的精度差异（即，纬度/经度：1.23/4.56789 可以解释为 1.23000 / 4.56780）。

  总体的想法是交错各个点，以最重要的数字按照从左到右的顺序排列。Accumulo 对地理位置按照词典字节序进行排列存储。这意味着，相同地理区域的点会相邻排列，以便有效地扫描。我们可以通过将查询键开始和结束参数分别使用 Z 曲线表示，实现快速地查找给定的地理区域。举个例子，查找在纬度/经度：30.1/60.2 和 40.87/0.9 之间的所有点，将会生成 120.1/150.2 和 130.8/160.9（偏移加 90）。这样 Z 曲线将生成一个查询的下界（开始）值 11250012 和一个上界（结束）值 11360089。这就是为什么对纬度/经度点上的整数部分进行补零操作是至关重要的。如果不做这个操作，应用程序将会把 1.23 放置在表的 10.3 附近，因为 Z 曲线进行洗牌的时候会为这个两个点都产生以 1 开始的行键。

地理空间位置只是我们 rowID 的一部分。当保存 ACLED 事件数据时，我们更喜欢将相同区域的事件按照时间逆序排列。GetReverseTime()方法通过在由 Z 曲线给定的条目后加上反向时间戳并用下划线隔开实现上述功能。这使我们可以在相同的 Accumulo 表中再加上时间范围限制查询（比如最近 100 条、最近 3 个月等）。具有相同纬度/经度的记录在 Accumulo 表中将按照升序排列，但最近的事件时间如果转换为毫秒精度后将变成一个很大的值。为了解决这个问题，我们将这个值减去一个最大值。如果传入的日志不匹配 yyyy-MM-dd 日期格式，我们将抛出一个异常。

最终的键格式是：Z 曲线点_反向时间戳。

单元测试主要是设计来测试在 `getZOrderCurve()` 方法和 `getReverseTime()` 方法中处理的错误，以及无效的输出结果。在将这些 ACLED 事件记录加载到 Accumulo 表之前，我们运行这个测试框架对我们的 rowID 生成器进行压力测试。

更多参考

本节所列出的 rowID 生成策略是用来适应经纬地理位置约束的，时间作为可选项限制查询事件。这很自由，当谈到 BigTable 的列存储设计时，并没有万能的 rowID 设计方案。根据你查询的类型的不同，你的 rowID 的生成策略可能完全不同。以下内容将进一步拓展本节的设计选择。

字典排序

Accumulo 设置存储在表中的键值对，按照键值进行字典排序。这意味着键值将按照对应的字节值排序，而不是预期的自然排序模式。例如，假设我们保存{1,2,10}序列作为 rowID，按照字典排序 10 将排在 1 之后，但是在 2 之前，这并不是我们期望的序列。本节使用补零点的办法绕开这个限制，创建一个固定长度的字符串来表示，这样按照字典排序和自然排序结果是一致的。补零后上述的序列将生成 01,02,10，当按照字典排序将保存序列顺序为 01,02,10。

这个技术在之前的章节中发挥了重要的作用。如果不按照固定长度方式，Z 曲线洗牌点 1.23,9.88 和 10.23,9.88 会使它们的顺序更接近数据空间的整体排序而不是技术序列。Z 曲线将分别生成 192838 和 19082830，这边给出了两个看上去不一致却在一起的两个点。在本节，偏移量设置为加 90，这样没有点可以超过 180，意味着整数部分最大只能是三位数。由零填充每个整数部分使其达到三位数（001.23 代替 1.23，010.23 代替 10.23，依此类推），rowID 从左到右的数字序列能更准确地反应出点的分布。

Z 曲线

Z 曲线是一种生成四叉树的技术，它代表一个扁平的地理空间数据的二维视图。更深入的信息可以通过维基百科 http://en.wikipedia.org/wiki/Z-order_curve 进行了解。

具体来说，本节生成 rowID 的策略对于区间查询纬度、经度地理点是非常灵活，其中边界参数的上下界的精度可以有所不同。rowID 按照从左到右显示数字意味着使用一个比较短的 Z 曲线查询模式比更长查询模式词可以匹配的结果会更多。举个例子，纬度/经度限制在 30.1/40.2 和 50.7/60.8，当进行交错时，这将生成起始键 340012 和结束键 560078。然而，同一个表可以用于更精确的查询范围比如 30.123/40.234 和 50.789/60.891，这样生成的查询键值是 3400122334 和 5600788991。前者，更少位数的查询起始键和结束键返回的结果比后者更多。

延伸阅读

❑ 使用 MapReduce 批量导入地理事件数据到 Accumulo（10.3 节）。

10.3 使用 MapReduce 批量导入地理事件数据到 Accumulo

本节将使用 MapReduce 加载以制表符分隔的 ACLED 事件数据到 Accumulo 库中。

准备工作

本节内容在伪分布式 Hadoop 集群下进行测试将是最容易的。测试环境为 Accumulo 1.4.1 和 Zookeeper 3.3.3。本章假定 Zookeeper 运行在本机的 2181 端口上，可以根据你的实际环境进行相应的更改。Accumulo 安装的 `bin` 目录需要配置到环境路径下。

在本节，我们将创建一个 `test` Accumulo 实例，设置相应的用户和密码。

你需要将数据集 `ACLED_nigeria_cleaned.tsv` 加载到 HDFS 的 `/input/acled_cleaned/` 路径下。

强烈推荐先阅读 10.1 节，本节将使用 `AccumuloTableAssistant.java` 和 `ACLEDRowIDGenerator.java` 类以及相应的父接口 `RowIDGenerator.java` 进行准备工作。

操作步骤

执行以下几个步骤，完成使用 MapReduce 批量加载事件数据到 Accumulo。

1. 打开你使用的 Java IDE。

2. 创建一个 build 模板用于生成一个名为 `accumulo_example.jar` 的 JAR 文件。

3. 创建包 `example.accumulo`,并添加 `RowIDGenerator.java`、`AccumuloTableAssistant.java` 和 `ACLEDRowIDGenerator.java`。

4. 你需要配置 Accumulo core 和 Hadoop 的 classpath 依赖。

5. 创建 `ACLEDIngest.java` 类:

```java
package examples.accumulo;

import org.apache.accumulo.core.client.mapreduce.AccumuloFileOutputFormat;
import org.apache.accumulo.core.client.mapreduce.lib.partition.RangePartitioner;
import org.apache.accumulo.core.data.Key;
import org.apache.accumulo.core.data.Value;
import org.apache.accumulo.core.util.CachedConfiguration;
import org.apache.hadoop.conf.Configuration;
import org.apache.hadoop.conf.Configured;
import org.apache.hadoop.fs.FileSystem;
import org.apache.hadoop.fs.Path;
import org.apache.hadoop.io.LongWritable;
import org.apache.hadoop.io.Text;
import org.apache.hadoop.mapreduce.Job;
import org.apache.hadoop.mapreduce.Mapper;
import org.apache.hadoop.mapreduce.Reducer;
import org.apache.hadoop.mapreduce.lib.input.FileInputFormat;
import org.apache.hadoop.mapreduce.lib.input.TextInputFormat;
import org.apache.hadoop.util.GenericOptionsParser;
import org.apache.hadoop.util.Tool;
import org.apache.hadoop.util.ToolRunner;

import java.io.IOException;
import java.util.regex.Pattern;

public class ACLEDIngest extends Configured implements Tool {

    private Configuration conf;

    public ACLEDIngest(Configuration conf) {
        this.conf = conf;
    }
```

6. 在 `run()` 方法中我们创建和提交作业。

```java
    @Override
    public int run(String[] args) throws Exception {

        if(args.length < 8) {
            System.err.println(printUsage());
```

使用 Apache Accumulo 进行持久化

```java
        System.exit(0);
    }

    Job job = new Job(conf, "ACLED ingest to Accumulo");
    job.setInputFormatClass(TextInputFormat.class);
    job.setMapperClass(ACLEDIngestMapper.class);
    job.setMapOutputKeyClass(Text.class);
    job.setMapOutputValueClass(Text.class);
    job.setReducerClass(ACLEDIngestReducer.class);
    job.setPartitionerClass(RangePartitioner.class);
    job.setJarByClass(getClass());

    String input = args[0];
    String outputStr = args[1];
    String instanceName = args[2];
    String tableName = args[3];
    String user = args[4];
    String pass = args[5];
    String zooQuorum = args[6];
    String localSplitFile = args[7];

    FileInputFormat.addInputPath(job, new Path(input));
    AccumuloFileOutputFormat.setOutputPath(job,clearOutputDir
(outputStr));
    job.setOutputFormatClass(AccumuloFileOutputFormat.class);
```

7. 创建 `AccumuloTableAssistant` 实例帮助创建和预分割 acled 表。

```java
    AccumuloTableAssistant tableAssistant = new AccumuloTableAssistant.Builder()
        .setInstanceName(instanceName)
        .setTableName(tableName).setUser(user)
        .setPassword(pass)
        .setZooQuorum(zooQuorum)
        .build();

    String splitFileInHDFS = "/tmp/splits.txt";
    int numSplits = 0;
    tableAssistant.createTableIfNotExists();
    if(localSplitFile != null) {
        numSplits = tableAssistant.
presplitAndWriteHDFSFile(conf, localSplitFile, splitFileInHDFS);
    }
    RangePartitioner.setSplitFile(job, splitFileInHDFS);
    job.setNumReduceTasks(numSplits + 1);

    if(job.waitForCompletion(true)) {
        tableAssistant.loadImportDirectory(conf, outputStr);
    }
    return 0;
}
```

8. 创建 `printUsage()` 和 `clearOutputDir()` 用于打印参数顺序以及清理提供的输出路径：

```java
    private String printUsage() {
        return "<input> <output> <instance_name> <tablename> +
                "<username> <password> <zoohosts> <splits_file_path>";
    }

    private Path clearOutputDir(String outputStr)throws IOException {
        FileSystem fs = FileSystem.get(conf);
        Path path = new Path(outputStr);
        fs.delete(path, true);
        return path;
    }
```

9. 创建内部 map 类 ACLEDIngestMapper.java。

```java
    public static class ACLEDIngestMapper
            extends Mapper<LongWritable, Text, Text, Text> {

        private Text outKey = new Text();
        private static final Pattern tabPattern = Pattern.compile("[\\t]");
        private ACLEDRowIDGenerator gen = new ACLEDRowIDGenerator();

        protected void map(LongWritable key, Text value,
                          Context context) throws IOException, InterruptedException {
            String[] values = tabPattern.split(value.toString());
            if(values.length == 8) {
                String [] rowKeyFields = new String[]
                // lat,lon,timestamp
                {values[4], values[5], values[1]};

                outKey.set(gen.getRowID(rowKeyFields));
                context.write(outKey, value);
            } else {
                context.getCounter("ACLED Ingest", "malformed records").increment(1l);
            }
        }
    }
```

10. 创建静态的 reduce 类 ACLEDIngestReducer.java:

```java
    public static class ACLEDIngestReducer
            extends Reducer<Text, Text, Key, Value> {

        private Key outKey;
        private Value outValue = new Value();
        private Text cf = new Text("cf");
        private Text qual = new Text();
        private static final Pattern tabPattern =
                           Pattern.compile("[\\t]");

        @Override
        protected void reduce(Text key, Iterable<Text> values,
```

```
                    Context context) throws IOException,
                InterruptedException {
        int found = 0;
        for(Text value : values) {
            String[] cells = tabPattern.split(value.toString());
            if(cells.length == 8) {
            // don't write duplicates
                if(found < 1) {
                    write(context, key, cells[3],"atr");
                    write(context, key, cells[1], "dtg");
                    write(context, key, cells[7], "fat");
                    write(context, key, cells[4], "lat");
                    write(context, key, cells[0], "loc");
                    write(context, key, cells[5], "lon");
                    write(context, key, cells[6], "src");
                    write(context, key, cells[2],"type");
                } else {
                    context.getCounter("ACLED Ingest",
                        "duplicates").increment(1l);
                }
            } else {
                context.getCounter("ACLED Ingest", "malformed records
missing a field").increment(1l);
            }
            found++;
        }
    }
}
```

11. 在 reduce 类中创建以下方法，用于生成输出的键值对：

```
        private void write(Context context, Text key, String cell,
                    String qualStr)
                throws IOException, InterruptedException {
            if(!cell.toUpperCase().equals("NULL")) {
                qual.set(qualStr);
                outKey = new Key(key, cf, qual, System.currentTimeMillis());
                outValue.set(cell.getBytes());
                context.write(outKey, outValue);
            }
        }
    }

    @Override
    public void setConf(Configuration conf) {
        this.conf = conf;
    }

    @Override
    public Configuration getConf() {
        return conf;
    }
```

12. 增加 main 类来提交你的作业实例给 `ToolRunner` 类：

    ```
    public static void main(String[] args) throws Exception {
        Configuration conf = CachedConfiguration.getInstance();
        args = new GenericOptionsParser(conf,
                                    args).getRemainingArgs();
        ToolRunner.run(new ACLEDIngest(conf), args);
    }
    }
    ```

13. 保存代码，编译 `accumulo_examples.jar` 到基本工作目录下。

14. 在工作目录下创建一个名为 `splits.txt` 文件，添加以下字符串：`00,01,10,11`。

15. 在工作目录下创建一个运行脚本 `bulk_ingest.sh`，内容如下：

    ```
    tool.sh accumulo_examples.jar examples.accumulo.ACLEDIngest\
    /input/acled_cleaned/\
    /output/accumulo_acled_load/\
    test\
    acled\
    root\
    password\
    localhost:2181\
    splits.txt
    ```

16. 运行该脚本你将会在 MapReduce 的 Web 界面上看到该作业的运行情况。完成后，可以扫描 Accumulo 中的 `acled` 表读取 ACLED 数据。

工作原理

该程序需要 8 个参数，每个都是非常重要的。输入路径是 ACLED 存储的位置。输出文件夹用于存储 Accumulo 原生的 Rfile 格式的输出数据。字符串 `test` 是本测试 Accumulo 实例在 Zookeeper 中存储的名称。字符串 `acled` 是我们希望创建的表名。我们通过使用字符串 `root:password` 进行权限认证。在本次执行中我们提供了一个 Zookeeper 节点 `localhost:2181`。最后，`splits.txt` 用于预分隔我们新创建的 `acled` 表。

本程序会清除输出路径下之前的所有数据。配置输出的格式化为 `AccumuloFileOutputFormat`。在这个作业中 map 输出结果的键和值都是 `Text`。

`AccumuloTableAssistant` 使用 Builder 模式进行链式调用以创建对象，避免在使用中出现参数错位。创建表 `acled`，并使用 `assistant` 根据提供的本地 `splits.txt` 文件对表进行预分片。如果在建表的时候不指定预分片，那么 `RangePartitioner` 将强制将中间结果键值对传递到一个单一的 reducer 上。根据预测的行键分布再创建预分片表将更有效，并且支持并行创建 RFiles。我们设置 reducer 的数量是 `splits.txt` 文件中实体数量加 1，用于处理那些落在最高分片点之上的键。最后,我们准备提交作业并观察 map 和 reduce 运行。

每个 map 任务的虚拟机都创建了一个 `ACLEDRowIDGenerator` 的实例。可以查看本章在 Accumulo 中设计行键存储地理事件这一节,深入理解这个类是如何工作的。我们的数据是按照制表符进行分隔的,并且遵循了严格的列排序,因此我们可以按照各自的顺序通过自定义列指数读取经纬度值和 `dtg`。键生成器需要这些字段生成一个有效的地理和倒排时间戳的复合 rowID。我们将生成的行键按照行输出为文本值。这样就生成了一个唯一的中间键用于插入 Accumulo 作为唯一的 rowID。

reducer 主要负责输出 rowID 和读取其他分隔符,输出一个等效的 rowID。rowID 生成器在 map 阶段基于 `lat`、`lon` 和 `dtg` 生成一个唯一的 rowID。按照定义,一个 ACLED 事件发生在同一个地点,同一个时间,将会被分组到同一个 reducer 中。但是,如果多个 ACLED 事件拥有相同的 rowID,意味着存在重复的条目,我们希望避免这种情况发生。因此我们在 reducer 中迭代对象的时候只保存第一个值。这个作业不做任何的去重处理。我们使用一个计数器来跟踪重复和没有分割正确的无效行。因为我们是直接将 Key/Value 实例写为 RFiles 的,Accumulo 要求 key/value 对象有序地插入。每个限定符的 rowID 自然是相同的,列簇也是一个静态标签 `cf`,但重要的是我们需要保持词典序的同时也要按照我们的限定标签考虑写顺序。

幸运的是,我们的数据是可预测的,我们硬编码列值,而列值读取是基于我们限定标签的字母排序。

当作业完成,所有的 RFiles 都是预分片好的,我们使用 assistant 实例读取所有生成的文件并直接输出存储在合适的分片上。这些数据立即就可以在 Accumulo 的 `acled` 表中被查询到。

更多参考

下面更详细解释一下你在本节看到的设计。

AccumuloTableAssistant.java

这个类设计为可以在不同的 Accumulo 数据加载和管理应用程序之间进行重用。因为它需要输入五个字符串参数,Builder 模式是最优的选择,可以防止参数错位。可以查看 Joshua Block 写的 Effective Java 2.0 有关 Builder 设计模式的细节。

分割点

00、01、10 和 11 作为分割点的选择,是完全随机的。它强调了预分片在建表过程中的重要性。正确的分割点设置取决于 rowID 的分布。分割点设置过少将影响作业在 reduce

阶段的吞吐量。设置过多，可能会浪费资源和启动时间，不能充分利用 reduce 任务 JVM。

`AccumuloOutputFormat` 和 `AccumuloFileOutputFormat`

如果你的数据容量很大，`AccumuloFileOutputFormat` 是显而易见的选择。生成 Rfile 直接插入分片库中不受 `AccumuloOutputFormat` 直接写 Accumulo 库的开销限制。另一方面，如果你的作业不是写密集型的，使用 Mutation 实例替代 Rfile 将更容易些。此外，如果你的作业是一个只有 map 的作业，`AccumuloOutputFormat` 和直接写 mutation 将是一个非常简单的设计选择。

延伸阅读

- 在 Accumulo 中设计行键存储地理事件（10.2 节）

10.4 设置自定义字段约束 Accumulo 中的地理事件数据

在本节，我们将编译一个自定义的 `Constraint` 类来限制 mutation 的类型，我们可以将其应用到 Accumulo 表的事件日期值中。具体地，我们期望新输入的值与一个特定的 `SimpleDateFormat` 模式匹配。但是这些值是表服务器的系统时间，不应该是在未来的某个时间。

准备工作

本节的内容将很容易在装有 Accumulo 1.4.1 版本和 Zookeeper 3.3.3 版本的伪分布式 Hadoop 集群上进行测试。本节的 shell 脚本假定了 Zookeeper 服务是启动在本机的 `2181` 端口，你可以根据你的环境更改主机名和端口。Accumulo 安装目录下的 `bin` 目录需要配置到你的环境路径下。

在本节，你需要使用用户名为 `root` 的账户和密码 `password` 创建一个名为 `test` 的 Accumulo 实例。

你需要在 Accumulo 实例中创建一个名为 `acled` 的表。

推荐你参考一下 10.3 节，这会给你提供一些样本数据。

操作步骤

完成以下步骤在 Accumulo 中实现和完成一个约束：

1. 打开你的 Java IDE，你需要配置 Accumulo core 和 Hadoop 的依赖：
2. 创建一个编译脚本用于生成一个名为 `accumulo_examples.jar` 的 JAR 文件。

3. 创建包 example.accumulo 和类 DtgConstraint.java 内容如下：

```java
package examples.accumulo;

import org.apache.accumulo.core.constraints.Constraint;
import org.apache.accumulo.core.data.ColumnUpdate;
import org.apache.accumulo.core.data.Mutation;

import java.text.DateFormat;
import java.text.ParseException;
import java.text.SimpleDateFormat;
import java.util.ArrayList;
import java.util.List;

public class DtgConstraint implements Constraint {

    private static final short DATE_IN_FUTURE = 1;
    private static final short MALFORMED_DATE = 2;
    private static final byte[] dtgBytes = "dtg".getBytes();
    private static final DateFormat dateFormatter = new
                    SimpleDateFormat("yyyy-MM-dd");

    public String getViolationDescription(short violationCode) {
        if(violationCode == DATE_IN_FUTURE) {
            return "Date cannot be in future";
        } else if(violationCode == MALFORMED_DATE) {
            return "Date does not match simple date format yyyy-MM-dd";
        }
        return null;
    }
```

4. 实现 check() 方法。

```java
    @Override
    public List<Short> check(Environment env, Mutation mutation) {
        List<Short> violations = null;
        try {
            for(ColumnUpdate update : mutation.getUpdates()) {
                if(isDtg(update)) {
                    long dtgTime = dateFormatter.parse(new
                      String(update.getValue())).getTime();
                    long currentMillis = System.currentTimeMillis();
                    if(currentMillis < dtgTime) {
                        violations = checkAndAdd(violations, DATE_IN_FUTURE);
                    }
                }
            }
        } catch (ParseException e) {
            violations = checkAndAdd(violations,MALFORMED_DATE);
        }
        return violations;
    }
}
```

5. 做一个字节的比较来检查更新是否符合 dtg 效验:

```java
    private boolean isDtg(ColumnUpdate update) {
        byte[] qual = update.getColumnQualifier();
        if(qual.length != dtgBytes.length)
            return false;
        for (int i = 0; i < qual.length; i++) {
            if(!(qual[i] == dtgBytes[i])) {
                return false;
            }
        }
        return true;
    }

    private List<Short> checkAndAdd(List<Short> violations,
                                    short violationCode) {
        if(violations == null)
            violations = new ArrayList<Short>();
        violations.add(violationCode);
        return violations;
    }
}
```

6. 保存这个类。

7. 在相同的包 examples.accumulo 中创建 DtgConstraintMain.java,实现代码如下:

```java
package examples.accumulo;

import org.apache.accumulo.core.client.*;
import org.apache.accumulo.core.conf.Property;
import org.apache.accumulo.core.data.ConstraintViolationSummary;
import org.apache.accumulo.core.data.Mutation;
import org.apache.accumulo.core.data.Value;
import org.apache.hadoop.io.Text;

import java.util.List;

public class DtgConstraintMain {
    public static final long MAX_MEMORY= 10000L;
    public static final long MAX_LATENCY=1000L;
    public static final int MAX_WRITE_THREADS = 4;
    public static final String TEST_TABLE = "acled";
    public static final Text COLUMN_FAMILY = new Text("cf");
    public static final Text DTG_QUAL = new Text("dtg");
```

8. 主函数对有效和无效的 dtg 值都进行插入尝试,用于测试我们的约束:

```java
    public static void main(String[] args) throws Exception {
        if(args.length < 6) {
System.err.println("examples.accumulo.DtgConstraintMain <row_id> <dtg> <instance_name> <user> <password> <zookeepers>");
```

```java
            System.exit(0);
        }
        String rowID = args[0];
        byte[] dtg = args[1].getBytes();
        String instanceName = args[2];
        String user = args[3];
        String pass = args[4];
        String zooQuorum = args[5];
        ZooKeeperInstance ins;
        Connector connector = null;
        BatchWriter writer = null;
        try {

            ins = new ZooKeeperInstance(instanceName,zooQuorum);
            connector = ins.getConnector(user, pass);
            writer = connector.createBatchWriter(TEST_TABLE, MAX_ MEMORY,
                            MAX_LATENCY, MAX_WRITE_THREADS);
connector.tableOperations().setProperty(TEST_TABLE, Property.
TABLE_CONSTRAINT_PREFIX.getKey() + 1, DtgConstraint.class.
getName());
            Mutation validMutation = new Mutation(new Text(rowID));
            validMutation.put(COLUMN_FAMILY, DTG_QUAL, new Value(dtg));
            writer.addMutation(validMutation);
            writer.close();
        } catch (MutationsRejectedException e) {
            List<ConstraintViolationSummary> summaries =
                e.getConstraintViolationSummaries();
            for (ConstraintViolationSummary sum : summaries) {
                System.err.println(sum.toString());
            }
        }
    }
}
```

9. 编译 JAR 文件，生成 `accumulo_examples.jar`。

10. 切换至 Accumulo 安装的本地目录，`$ACCUMULO_HOME/conf`，编辑 `accumulo-site.xml` 文件。

11. 编辑 `accumulo-site.xml` 配置文件的 `general.classpaths` 属性，让其包含 `accumulo_examples.jar` 的路径。

12. 使用 `$ACCUMULO_HOME/bin` 目录下的 `tdown.sh` 和 `tup.sh` 重启 TabletServer。

13. 使用以下命令来测试是否在 Accumulo 路径中包含了该 JAR 文件：

`$ accumulo classpath`

你将看见一个文件打印输出包含 `accumulo_examples.jar`。

14. 将 `accumulo_examples.jar` 文件所在的目录做为工作根目录,在该目录下创建一个名为 `run_constraint_test.sh` 的 shell 脚本,并编辑为下面的命令行。请确认修改的 `ACCUMULO-LIB`、`HADOOP_LIB` 和 `ZOOKEEPER_LIB` 与你本地的环境相匹配:

```
ACCUMULO_LIB=/opt/cloud/accumulo-1.4.1/lib/*
HADOOP_LIB=/Applications/hadoop-0.20.2-cdh3u1/*:/Applications/hadoop-0.20.2-cdh3u1/lib/*
ZOOKEEPER_LIB=/opt/cloud/zookeeper-3.4.2/*
java -cp $ACCUMULO_LIB:$HADOOP_LIB:$ZOOKEEPER_LIB:accumuloexamples.jar examples.accumulo.DtgConstraintMain\
00993877573819_9223370801921575807\
2012-08-07\
test\
root\
password\
localhost:2181
```

15. 保存并运行脚本,它应该悄悄地完成。

16. 编辑 `run_constraint_test.sh` 脚本,修改 `dtg` 参数为 2012-08-07 到 2030-08-07。

17. 保存并再次运行该脚本。你将会看到在控制台打印出一个约束错误指示 `Date cannot be in future`。

工作原理

我们的约束类遍历每一个 mutation,确定列限定符是否匹配相关的 `dtg`。如果列更新的键值对包含了 `dtg` 限定符,那么表明该值是错误的。有如下两个约束条件。

1. 时间并不匹配 Java 时间模式 yyyy-MM-dd。因此 1970-12-23 和 2013-02-11 将通过,70-12-23 或者 12-20-22 将生成一个错误并添加一条违反约束。

2. 写入那一刻是未来的一个时间点,如 2030-08-07 是 18 年后的一个时间。如果某一列更新包含了一个未来的日期,将添加一条违反约束。

主函数将所需的参数传递到 Accumulo 实例中,并且添加约束给具体的表。然后尝试在提供的 rowID 使用参数提供的 `dtg` 进行 mutation。如果 mutation 因一些理由被拒绝,打印出违反约束,确认是否违反了 `dtg` 约束。

我们可以通过修改 shell 脚本的 `dtg` 参数,观察产生的不同约束违反错误。

更多参考

约束是 Accumulo 数据库执行数据策略的一个强大的功能。下面将讨论一些其他你应该知道的事情。

内置约束类

Accumulo 的核心包提供了众多非常好的约束实现类。涵盖了各种常见的条件检查，并且都已经包含在 TabletServer 的类路径下。可在 `org.apache.accumulo.examples.simple.constraints` 包下 `simple` 例子模块中查看实现范例，也可在 Accumulo 的其他核心系统内部使用约束实现的场景中查看。

在每个 TabletServer 上安装一个约束

如果在你的 Accumulo 中安装一个自定义的约束后，你注意到每一个 mutation 都被拒绝；无论是什么原因，很可能该 TabletServer 服务器的类路径中找不到你的约束类。查看 TabletServer 的服务器日志，发现有 `ClassNotFoundExceptions`。这可能发生在表配置了约束类列表，但却找不到一个名称完全匹配的类的情况下。在一个完全分布式的环境下，修改完类路径以后，要确认重启了所有的 TabletServer。

延伸阅读

- 使用 MapReduce 批量导入地理事件数据到 Accumulo（10.3 节）。
- 使用 Accumulo 实行单元级安全的扫描（10.7 节）。

10.5 使用正则过滤器限制查询结果

本章将使用 Accumulo 内置的 `RegExFilter` 返回一个特定限定符的键值对。这些过滤器将分发到存有 `acled` 表的不同服务器上。

准备工作

本节的内容将很容易在装有 Accumulo 1.4.1 版本和 Zookeeper 3.3.3 版本的伪分布式 Hadoop 集群上进行测试。本节的 shell 脚本假定了 Zookeeper 服务是启动在本机的 2181 端口，你可以根据你的环境更改主机名和端口。

在本节，你需要使用名为 `root` 的账户和密码 `password` 创建一个名为 `test` 的 Accumulo 实例。

为了看到本节过滤的结果，你需要先完成 10.3 节。这可以为我们创造一些可试验的样例数据。

操作步骤

执行以下步骤使用正则过滤迭代器。

1. 打开你的 Java IDE，你需要在 classpath 中配置 Accumulo 核心包和 Hadoop 的依赖包。

2. 创建一个 build 模板，生成名为 accumulo_examples.jar 的文件。

3. 创建包 example.accumulo 并添加类 SourceFilterMain.java，添加如下内容：

```java
package examples.accumulo;

import org.apache.accumulo.core.client.Connector;
import org.apache.accumulo.core.client.IteratorSetting;
import org.apache.accumulo.core.client.Scanner;
import org.apache.accumulo.core.client.ZooKeeperInstance;
import org.apache.accumulo.core.data.Key;
import org.apache.accumulo.core.data.Value;
import org.apache.accumulo.core.iterators.user.RegExFilter;
import org.apache.accumulo.core.security.Authorizations;
import org.apache.hadoop.io.Text;

import java.util.HashMap;
import java.util.Map;

public class SourceFilterMain {

    public static final String TEST_TABLE = "acled";

    public static final Text COLUMN_FAMILY = new Text("cf");
    public static final Text SRC_QUAL = new Text("src");
```

4. 主函数处理参数解析并使用过滤器进行查询：

```java
    public static void main(String[] args) throws Exception {
        if(args.length < 5) {
            System.err.println("usage: <src> <instance name> <user> <password> <zookeepers>");
            System.exit(0);
        }
        String src = args[0];
        String instanceName = args[1];
        String user = args[2];
        String pass = args[3];
        String zooQuorum = args[4];
        ZooKeeperInstance ins = new
                ZooKeeperInstance(instanceName, zooQuorum);
        Connector connector = ins.getConnector(user, pass);
        Scanner scan = connector.createScanner(TEST_TABLE, new Authorizations());
        scan.fetchColumn(COLUMN_FAMILY, SRC_QUAL);
        IteratorSetting iter = new IteratorSetting(15, "regexfilter", RegExFilter.class);
        iter.addOption(RegExFilter.VALUE_REGEX, src);
        scan.addScanIterator(iter);
        int count = 0;
        for(Map.Entry<Key, Value> row : scan) {
```

```
            System.out.println("row: " +
                        row.getKey().getRow().toString());
            count++;
        }
        System.out.println("total rows: " + count);
    }
}
```

5. 保存并编译 JAR 文件 `accumulo_examples.jar`。

6. 将 `accumulo_examples.jar` 文件所在的目录做为工作根目录，在该目录下创建一个名为 `run_src_filter.sh` 的 shell 脚本，并编辑为下面的命令行。请确认修改的 `ACCUMULO-LIB`、`HADOOP_LIB` 和 `ZOOKEEPER_LIB` 与你本地的环境相匹配。

```
ACCUMULO_LIB=/opt/cloud/accumulo-1.4.1/lib/*
HADOOP_LIB=/Applications/hadoop-0.20.2-cdh3u1/*:/Applications/hadoop-0.2
0.2-cdh3u1/lib/*
ZOOKEEPER_LIB=/opt/cloud/zookeeper-3.4.2/*
java -cp $ACCUMULO_LIB:$HADOOP_LIB:$ZOOKEEPER_LIB:accumuloexamples.
jar examples.accumulo.SourceFilterMain\
 'Panafrican News Agency'\
 test\
 root\
 password\
 localhost:2181
```

7. 保存并运行脚本。你将会看到为过滤源 `Panafrican News Agency` 返回了 49 行结果。

工作原理

该脚本需要输入一个必要的参数用于连接 Accumulo 的 `acled` 表，同时还需要一个附加参数作为源限定值来构造过滤器。我们建立一个无认证的扫描器，该扫描器配置了一个类型为 `RegExFilter` 的 `IteratorSetting`，用于在表服务器上做正则比较。我们的正则比较非常简单，直接匹配提供的源参数。

然后我们遍历结果集并打印出所有匹配键值对的 rowID。最后我们打印出一共有多少行记录被匹配。

基于值比较的过滤器需要将键值对分发到许多拥有 `acled` 表的 TabletServers 服务器上。客户端只会看到那些被匹配到的行，并可以直接进行处理。

延伸阅读

- 使用 MapReduce 批量导入地理事件数据到 Accumulo（10.3 节）。
- 使用 Accumulo 实行单元级安全的扫描（10.7 节）。

10.6 使用 SumCombiner 计算同一个键的不同版本的死亡数总和

本节使用 Accumulo 内置的 SumCombiner，对表 `acled` 的每一个键，将与限定符 `fat` 关联的单元值看做 `long` 型，对这个键的所有版本的对应值进行求和。

准备工作

在安装了 Accumulo 1.4.1 以及 Zookeeper 3.3.3 的伪分布式 Hadoop 集群上进行本节的测试是最容易的。本节的 shell 脚本假定 Zookeeper 运行的主机为 `localhost`，端口号为 `2181`。你可以修改这些值从而满足你自身环境的需要。Accumulo 安装后的 `bin` 文件夹需要设置为你的环境路径。

对于本节，需要创建一个名为 `test` 的 Accumulo 实例，用户为 `root`，对应的密码为 `password`。

需要在已配置好的 Accumulo 实例创建一张表 `acled`。如果在以前的章节已经建立了这样一张表，需要删除并重建该表。

我们强烈建议你完成在 10.3 节中的脚本。这将提供用于下面实验的若干采样数据。

操作步骤

下面的步骤讨论了使用 SumCombiner 的查询。

1. 选择打开一个 Java IDE，配置 Accumulo 以及 Hadoop 的路径依赖。

2. 创建一个 build 模板，生成一个 JAR 文件，命名为 accumulo_examples.jar。

3. 创建一个包 `example.accumulo` 并加入包含如下内容的 TotalFatalityCombinerMain.java 类：

```java
package examples.accumulo;

import org.apache.accumulo.core.client.*;
import org.apache.accumulo.core.client.Scanner;
import org.apache.accumulo.core.data.*;
import org.apache.accumulo.core.iterators.Combiner;
import org.apache.accumulo.core.iterators.LongCombiner;
import org.apache.accumulo.core.iterators.user.SummingCombiner;
import org.apache.accumulo.core.security.Authorizations;
import org.apache.hadoop.io.Text;
```

```
import java.util.*;

public class TotalFatalityCombinerMain {

    public static final long MAX_MEMORY= 10000L;
    public static final long MAX_LATENCY=1000L;
    public static final int MAX_WRITE_THREADS = 4;
    public static final String TEST_TABLE = "acled";
    public static final Text COLUMN_FAMILY = new Text("cf");
    public static final Text FATALITIES_QUAL = new Text("fat");
```

4. main()方法处理参数的解析:

```
    public static void main(String[] args) throws Exception {
        if(args.length < 4) {
            System.err.println("usage: <instance name>
                    <user> <password> <zookeepers>");
            System.exit(0);
        }
        String instanceName = args[0];
        String user = args[1];
        String pass = args[2];
        String zooQuorum = args[3];
        ZooKeeperInstance ins = new
            ZooKeeperInstance(instanceName, zooQuorum);
        Connector connector = ins.getConnector(user, pass);
        if(!connector.tableOperations().exists(TEST_TABLE))
            connector.tableOperations().create(TEST_TABLE);
        BatchWriter writer = connector.createBatchWriter(TEST_TABLE,
MAX_MEMORY, MAX_LATENCY, MAX_WRITE_THREADS);
```

5. 写入一些相同 rowID eventA、列族以及限定符的采样数据:

```
        Mutation m1 = new Mutation("eventA");
        m1.put(COLUMN_FAMILY, FATALITIES_QUAL, new
            Value("10".getBytes()));

        Mutation m2 = new Mutation("eventA");
        m2.put(COLUMN_FAMILY, FATALITIES_QUAL, new
            Value("5".getBytes()));
```

6. 写入另外一个以 rowID eventB 为键的数据:

```
        Mutation m3 = new Mutation("eventB");
        m3.put(COLUMN_FAMILY, FATALITIES_QUAL, new
            Value("7".getBytes()));

        writer.addMutation(m1);
        writer.addMutation(m2);
```

```
            writer.addMutation(m3);
            writer.close();
```

7. 使用 combiner 为扫描器配置一个 `IteratorSetting`：

```
            IteratorSetting iter = new IteratorSetting(1,
                            SummingCombiner.class);
            LongCombiner.setEncodingType(iter,
                            SummingCombiner.Type.STRING);
            Combiner.setColumns(iter,
                    Collections.singletonList(new
                        IteratorSetting.Column(COLUMN_FAMILY,
                                FATALITIES_QUAL)));
            Scanner scan = connector.createScanner(TEST_TABLE,
                    new Authorizations());
            scan.setRange(new Range(new Text("eventA"), new
                            Text("eventB")));
            scan.fetchColumn(COLUMN_FAMILY, FATALITIES_QUAL);
            scan.addScanIterator(iter);
            for(Map.Entry<Key, Value> item : scan) {
                System.out.print(item.getKey().getRow().toString() +
":             fatalities: ");
                System.out.println(new
                            String(item.getValue().get()));
            }
        }
    }
```

8. 保存并编译 JAR 文件 `accumulo_examples.jar`。

9. 在 `accumulo_examples.jar` 所处的工作路径中，创建一个新的名为 `run_combiner.sh` 的 shell 脚本，其中包含下面的命令。需要确认的是将 `ACCUMULO-LIB`、`HADOOP_LIB` 以及 `ZOOKEEPER_LIB` 修改成与你的本地路径一致：

```
ACCUMULO_LIB=/opt/cloud/accumulo-1.4.1/lib/*
HADOOP_LIB=/Applications/hadoop-0.20.2-cdh3u1/*:/Applications/hadoop-0.20.2-cdh3u1/lib/*
ZOOKEEPER_LIB=/opt/cloud/zookeeper-3.4.2/*
java -cp $ACCUMULO_LIB:$HADOOP_LIB:$ZOOKEEPER_LIB:accumulo-examples.jar examples.accumulo.TotalTatality Combiner Main\
test\
root\
password\
localhost:2181
```

10. 保存并运行脚本。

11. 在命令行会看见如下输出结果：

```
eventA: fatalities: 15
eventB: fatalities: 7
```

12. 重新运行脚本。

13. 现在你将看见对于每个事件都记数了两次：

```
eventA: fatalities: 30
eventB: fatalities: 14
```

工作原理

类 `TotalFatalityCombinerMain` 读取连接 Accumulo 以及实例化写测试数据至 `acled` 表的实例 `BatchWriter` 的参数。我们记录了两个不同的版本但包含相同 rowID 为 `eventA` 的键的记录。一个包含的限定符 `fat` 的值为 10，另外一个则为 5，另外，我们记录了一个包含 rowID 为 `eventB` 的键的记录，并且限定符 `fat` 的值为 7。

接下来，使用 `Scanner` 实例，应用 `SumCombiner` 对表中的键值对扫描。combiner 的任务是收集对于相同键关联的不同 `long` 型值，并输出这些 `long` 型值的和。值 5 以及 10 与同样的键（rowID `eventA`）关联，合并之后得到结果 15。另外，只有一个版本值与 rowID 为 `eventB` 的键关联，因此仍然将单个值 7 作为这个键的和返回。

如果我们重新运行该应用，前面的那些记录仍然存储在同样的 Accumulo 表中。重新运行该应用再次提供了相同的记录。对于 rowID `eventA` 增加了值 5 和 10，对于 rowID `eventB` 增加了值 7。

现在重新运行 Combiner 扫描器，可知对于 rowID `eventA`，包含 4 个条目（5,10,5,10）；同样的对于 rowID `eventB`，包含 2 个条目（7,7）。计算的结果是前面结果的两倍。如果不清除表中数据，每次运行该应用，得到的结果都会增加 15 以及 7。

会发生这样情况的原因是，在原始默认的键值存储级别下，当应用调用一次就会将包含不同时间戳的记录以新的键值对的形式插入表中。我们的 combiner 会识别不同键的所有时间戳版本记录。

更多参考

这里有一些关于 combiner 的帮助信息。

Combiner 是基于每个键的，而非所有的键

对于使用 Accumulo 的新用户，会产生疑惑。Combiner 使用对键值聚合的 Accumulo 迭代器模式，只能够基于每个键的不同版本的记录。如果你需要对一个限定符的值进行全表

范围的聚合，你仍然需要使用 MapReduce 完成操作。具体参见 10.8 节。

Combiner 能应用在扫描的时候，也可以通过表的配置应用于对新进记录进行合并

本节使用 combiner 在扫描的时候对限定符的值进行聚合。Accumulo 同样支持将 combiner 持久化至一个表的配置中，这样当有记录写入时就会对其合并。

延伸阅读

- 使用 MapReduce 批量的将地理事件数据导入 Accumulo（10.3 节）。
- 使用正则表达式过滤迭代器限制输出结果（10.5 节）。
- 使用 MapReduce 聚合 Accumulo 中的消息源（10.8 节）。

10.7 使用 Accumulo 实行单元级安全的扫描

Accumolo 为表中每个单独的键值对提供了应用单元级可见性标签的能力，这是区别于其他 BigTalbe 实现的最显著的特征。本节将展示一种应用单元级安全的方法。本节代码中的一些变量，只能通过适当的授权，进行扫描和读取操作。

准备工作

在安装了 Accumulo 1.4.1 以及 Zookeeper 3.3.3 的伪分布式 Hadoop 集群上进行本节的测试是最容易的。本节的 shell 脚本假定 Zookeeper 运行的主机为 `localhost`，端口号为 `2181`。你可以修改这些值从而满足你自身环境的需要。Accumulo 安装后的 `bin` 文件夹需要设置为你的环境路径。

对于本节，需要创建一个名为 `test` 的 Accumulo 实例，用户为 `root`，对应的密码为 `password`。

你需要在已配置好的 Accumulo 实例创建一张表 `acled`。如果在以前的章节已经建立了这样一张表，需要删除并重建该表。

为了得到通过本节代码过滤后的结果，需要执行在 10.3 节中的脚本。这将提供用于下面实验的若干采样数据。

操作步骤

执行下面的步骤，使用单元级控制读写 `Accumulo` 中数据。

使用 Apache Accumulo 进行持久化　233

1. 选择打开一个 Java IDE，配置 Accumulo 以及 Hadoop 的路径依赖。
2. 创建一个 build 模板，生成一个 JAR 文件，命名为 `accumulo_examples.jar`。
3. 创建一个包 `example.accumulo` 并加入包含如下内容的 `SecurityScanMain.java` 类：

```java
package examples.accumulo;

import org.apache.accumulo.core.client.*;
import org.apache.accumulo.core.data.Key;
import org.apache.accumulo.core.data.Mutation;
import org.apache.accumulo.core.data.Value;
import org.apache.accumulo.core.security.Authorizations;
import org.apache.accumulo.core.security.ColumnVisibility;
import org.apache.hadoop.io.Text;

import java.util.Map;

public class SecurityScanMain {

    public static final long MAX_MEMORY= 10000L;
    public static final long MAX_LATENCY=1000L;
    public static final int MAX_WRITE_THREADS = 4;
    public static final String TEST_TABLE = "acled";
    public static final Text COLUMN_FAMILY = new Text("cf");
    public static final Text THREAT_QUAL = new Text("trt_lvl");

    public static void main(String[] args)throws Exception {
        if(args.length < 4) {
            System.err.println("usage: <instance name> <user> <password> <zookeepers>");
            System.exit(0);
        }
        String instanceName = args[0];
        String user = args[1];
        String pass = args[2];
        String zooQuorum = args[3];
```

4. 创建一个 `Connector` 实例，将变量 `user` 以及 `pass` 传给 Accumulo 实例的 `test` 实例。

```java
        ZooKeeperInstance ins = new
                ZooKeeperInstance(instanceName, zooQuorum);
        Connector connector = ins.getConnector(user, pass);
        if(!connector.tableOperations().exists(TEST_TABLE))
            connector.tableOperations().create(TEST_TABLE);
```

5. 获得 `root` 用户的当前权限。

```java
        Authorizations allowedAuths =
connector.securityOperations().getUserAuthorizations(user);
```

```
            BatchWriter writer =
connector.createBatchWriter(TEST_TABLE, MAX_MEMORY,
                     MAX_LATENCY, MAX_WRITE_THREADS);
```

6. 编写测试例子。

```
            Mutation m1 = new Mutation(new Text("eventA"));
            m1.put(COLUMN_FAMILY,
                   THREAT_QUAL,
                   new ColumnVisibility("(p1|p2|p3)"),
                   new Value("moderate".getBytes()));
            Mutation m2 = new Mutation(new Text("eventB"));
            m2.put(COLUMN_FAMILY,
                   THREAT_QUAL,
                   new ColumnVisibility("(p4|p5)"),
                   new Value("severe".getBytes()));
            writer.addMutation(m1);
            writer.addMutation(m2);
            writer.close();
```

7. 使用当前用户权限创建一个 scanner，获取所有键中包含限定符 threat 的键值对。

```
            Scanner scanner = connector.createScanner(TEST_TABLE,
allowedAuths);
            scanner.fetchColumn(COLUMN_FAMILY, THREAT_QUAL);
            boolean found = false;
            for(Map.Entry<Key, Value> item: scanner) {
                System.out.println("Scan found: " + item.getKey().
getRow().toString() + " threat level: " + item.getValue().toString());
                found = true;
            }
```

8. 如果判断逻辑满足，表明当前用户是没有权限查看任何"威胁事件"的。

```
            if(!found)
                System.out.println("No threat levels are visible with
your current user auths: " + allowedAuths.serialize());
        }
}
```

9. 保存并编译 JAR 文件 accumulo_examples.jar。

10. 在 accumulo_examples.jar 所处的工作路径中，创建一个新的名为 run_security_auth_scan.sh 的 shell 脚本，其中包含下面的命令。需要确认的是将 ACCUMULO-LIB、HADOOP_LIB 及 ZOOKEEPER_LIB 修改成与你的本地路径一致：

```
ACCUMULO_LIB=/opt/cloud/accumulo-1.4.1/lib/*
HADOOP_LIB=/Applications/hadoop-0.20.2-cdh3u1/*:/Applications/
```

```
hadoop-0.20.2-cdh3u1/lib/*
ZOOKEEPER_LIB=/opt/cloud/zookeeper-3.4.2/*
java -cp $ACCUMULO_LIB:$HADOOP_LIB:$ZOOKEEPER_LIB:accumulo-
examples.jar examples.accumulo.SecurityScanMain\
 test\
 root\
 password\
 localhost:2181
```

11. 保存并运行脚本。

12. 在命令行会看见如下输出结果：

```
no threat levels are visible with your current user auths:
```

13. 启动 Accumulo shell。

```
accumulo shell -u root -p password
```

14. 执行 setauths 命令，会见到一个选项列表：

```
$ root@test> setauths
```

15. 运行下面的命令：

```
$ root@test> setauths -s p1
```

16. 再运行一次脚本 run_security_auth_scan.sh。

17. 在命令行会看见如下输出结果：

```
Scan found: eventA threat level: moderate
```

18. 在 Accumulo shell 中再次输入下面的命令：

```
$ root@test> setauths -s p1,p4
```

19. 再运行一次脚本 run_security_auth_scan.sh。

20. 在命令行会看见如下输出结果：

```
Scan found: eventA threat level: moderate
Scan found: eventB threat level: severe
```

工作原理

类 `SecurityScanMain` 读取参数连接 Accumulo 以及实例化写测试数据至 `acled` 表的实例 `BatchWriter` 的参数。我们将两个变量记录写入表。第一个对于 rowID 为 `eventA` 的列可见性表达式 `(p1|p2|p3)`。第二个对于 rowID 为 `eventB` 的列可见性表达式 `(p4|p5)`。列可见性表达式是非常简单的布尔表达式。在扫描一个 Accumulo 表之前，客户端需要提

供已连接用户的验证符。Accumulo 会将得到的验证符与每个键的列可见性标签进行对比，从而确定这个用户对于给定键值对的可见性。表达式(p1|p2|p3)表明扫描器能读取的键需要表示为 p1、p2 或者 p3 提供的 Authorizations 对象。默认的，root 用户不包含任何扫描权限验证符。连接器调用 getUserAuthorizations(user) 方法，此时不会返回任何权限验证符。为了查看 eventA，需要提供 p1、p2 或者 p3，这些都未提供给 root 用户。为了查看 eventB，需要提供 p4 或者 p5，root 用户也都不包含。如果进入 shell 为 root 用户添加 p1，我们会提供 p1 授权，从而能成功的与 eventA 匹配。一旦我们将 root 的扫描验证符设置为 p1 以及 p4，我们就能同时查看 eventA 和 eventB。

更多参考

单元级可见性是一个比你想象更复杂的特性。这里有一些关于 Accumulo 中单元级安全性的知识。

为未授权的扫描设定变量记录

授权验证符限制的是用户扫描过程中所能看到的，而不是在 matation 上写的脚本能访问哪些表。

对于大多数系统，这是默认的行为，并且是不好的。如果你希望在自己安装的 Accumulo 执行这个规则，可以添加 org.apache.accumulo.core.security.VisibilityConstraint 的实现类 Constraint，作为系统级约束。一旦在 Accumulo 安装过程中执行了此操作，用户将禁止写入包含列可见性标签的变量记录，从而自己也没有权限读取。

列的可见性是键的一部分

不同的键有可能包含同样的 rowID，列族以及限定符，但是却包含不同的 ColumnVisibility 标签。如果最新时间戳版本的键包含的 ColumnVisibility 键不允许当前扫描可见，用户会获得之前与列可见性限定符匹配的下一个老版本的键值，或者没有任何版本与之匹配，如果用户对任何版本都未获授权。

正常的扫描逻辑是扫描器返回指定键对应最近版本的键值对。单元级可见性系统根据额外的条件，调整了该逻辑。扫描器将会返回与提供的授权验证符匹配的给定键对应的，最近时间戳版本的结果。

支持更复杂的布尔表达式

对于 ColumnVisibility 布尔表达式，本节展示了两个非常简单的析取表达式的例子。

你可以使用更加复杂的表达式，如果你的应用需要的话。比如，(((A & B)|C) & D)会匹配提供了标签 D 并且要么提供了标签 C 要么同时提供标签 A 和标签 B 的授权。

延伸阅读

- 使用 MapReduce 批量导入地理事件数据到 Accumulo（10.3 节）。
- 设置自定义字段约束 Accumulo 中的地理事件数据（10.4 节）。

10.8 使用 MapReduce 聚集 Accumulo 中的消息源

本节，我们使用 MapReduce 以及类 `Accumulo Input Format` 计算存储在一张 Accumulo 表中的每个独立消息源出现的次数。

准备工作

在安装了 Accumulo 1.4.1 以及 Zookeeper 3.3.3 的伪分布式 Hadoop 集群上进行本节的测试是最容易的。本节的 shell 脚本假定 Zookeeper 运行的主机为 `localhost`，端口号为 `2181`。你可以修改这些值从而满足你自身环境的需要。Accumulo 安装后的 `bin` 文件夹需要设置为你的环境路径。

对于本节，需要创建一个名为 `test` 的 Accumulo 实例，用户为 `root`，对应的密码为 `password`。

你需要在已配置好的 Accumulo 实例创建一张表 `acled`。

为了得到通过本节代码过滤后的结果，需要执行在 10.3 节中的脚本。这将提供用于下面实验的若干采样数据。

操作步骤

执行下面的步骤，使用 MapReduce 计算不同消息源出现的次数。

1. 选择打开一个 Java IDE，配置 Accumulo 以及 Hadoop 的路径依赖。
2. 创建一个 build 模板，生成一个 JAR 文件，命名为 `accumulo_examples.jar`。
3. 创建一个包 `example.accumulo` 并加入包含如下内容的 `SourceCountJob.java` 类：

```
package examples.accumulo;

import org.apache.accumulo.core.client.mapreduce.
AccumuloInputFormat;
```

```java
import org.apache.accumulo.core.data.Key;
import org.apache.accumulo.core.data.Value;
import org.apache.accumulo.core.security.Authorizations;
import org.apache.accumulo.core.util.CachedConfiguration;
import org.apache.accumulo.core.util.Pair;
import org.apache.hadoop.conf.Configuration;
import org.apache.hadoop.conf.Configured;
import org.apache.hadoop.fs.FileSystem;
import org.apache.hadoop.fs.Path;
import org.apache.hadoop.io.IntWritable;
import org.apache.hadoop.io.Text;
import org.apache.hadoop.mapreduce.Job;
import org.apache.hadoop.mapreduce.Mapper;
import org.apache.hadoop.mapreduce.Reducer;
import org.apache.hadoop.mapreduce.lib.output.FileOutputFormat;
import org.apache.hadoop.mapreduce.lib.output.TextOutputFormat;
import org.apache.hadoop.util.GenericOptionsParser;
import org.apache.hadoop.util.Tool;
import org.apache.hadoop.util.ToolRunner;

import java.io.IOException;
import java.lang.Override;
import java.util.HashSet;
public class SourceCountJob extends Configured implements Tool {

    private Configuration conf;
    private static final Text FAMILY = new Text("cf");
    private static final Text SOURCE = new Text("src");

    public SourceCountJob(Configuration conf) {
        this.conf = conf;
    }
```

4. 添加 run() 方法，确定 Tool 接口以及命令行参数的解析。

```java
    @Override
    public int run(String[] args) throws Exception {

        args = new GenericOptionsParser(conf,
            args).getRemainingArgs();
        if(args.length < 6) {
            System.err.println(printUsage());
            System.exit(0);
        }

        String tableName = args[0];
```

```
            String outputStr = args[1];
            String instanceName = args[2];
            String user = args[3];
            String pass = args[4];
            String zooQuorum = args[5];
```

5. 配置 Accumulo 输入设置。

```
            AccumuloInputFormat.setInputInfo(conf, user, pass.
getBytes(), tableName, new Authorizations());
            AccumuloInputFormat.setZooKeeperInstance(conf,
instanceName, zooQuorum);
            HashSet<Pair<Text, Text>> columnsToFetch = new
HashSet<Pair<Text,Text>>();
            columnsToFetch.add(new Pair<Text, Text>(FAMILY, SOURCE));
            AccumuloInputFormat.fetchColumns(conf, columnsToFetch);
```

6. 设置作业、map/reduce 类、输出位置：

```
            Job job = new Job(conf, "Count distinct sources in ACLED");
            job.setInputFormatClass(AccumuloInputFormat.class);
            job.setMapperClass(ACLEDSourceMapper.class);
            job.setMapOutputKeyClass(Text.class);
            job.setMapOutputValueClass(IntWritable.class);
            job.setReducerClass(ACLEDSourceReducer.class);
            job.setCombinerClass(ACLEDSourceReducer.class);
            job.setJarByClass(getClass());
            job.setOutputFormatClass(TextOutputFormat.class);
            FileOutputFormat.setOutputPath(job,
                            clearOutputDir(outputStr));
            job.setNumReduceTasks(1);
            return job.waitForCompletion(true) ? 0 : 1;
        }

        private String printUsage() {
            return "<tablename> <output> <instance_name>
                <username> <password> <zoohosts>";
        }

        private Path clearOutputDir(String outputStr)
                throws IOException {
            FileSystem fs = FileSystem.get(conf);
            Path path = new Path(outputStr);
            fs.delete(path, true);
            return path;
        }
```

7. 添加静态内部类 ACLEDSourceMapper。

```java
public static class ACLEDSourceMapper
        extends Mapper<Key, Value, Text, IntWritable> {

    private Text outKey = new Text();
    private IntWritable outValue = new IntWritable(1);

    @Override
    protected void map(Key key, Value value, Context context)
                    throws IOException, Interrupted Exception {

        outKey.set(value.get());
        context.write(outKey, outValue);
    }
}
```

8. 添加内部静态类 ACLEDSourceReducer。

```java
public static class ACLEDSourceReducer
        extends Reducer<Text, IntWritable, Text, IntWritable> {

    private IntWritable outValue = new IntWritable();

    @Override
    protected void reduce(Text key, Iterable<IntWritable> values,
                    Context context) throws
                IOException, InterruptedException {

      int count = 0;
      for(IntWritable value : values) {
        count += value.get();
      }
      outValue.set(count);
      context.write(key, outValue);
    }
}

@Override
public void setConf(Configuration conf) {
    this.conf = conf;
}

@Override
```

```
    public Configuration getConf() {
        return conf;
    }
```

9. 定义 `main()` 方法，用于作为提交作业的一个 `Tool` 的实例。

```
    public static void main(String[] args) throws Exception {
        Configuration conf = CachedConfiguration.getInstance();
        args = new GenericOptionsParser(conf, args).getRemainingArgs();
        ToolRunner.run(new SourceCountJob(conf), args);
    }
}
```

10. 保存并构建 JAR 文件 `accumulo_examples.jar`。

11. 在 `accumulo_examples.jar` 所处的工作路径中，创建一个新的名为 `source_count.sh` 的 shell 脚本，其中包含下面的命令。需要确认的是将 `ACCUMULO-LIB`、`HADOOP_LIB` 以及 `ZOOKEEPER_LIB` 修改成与你的本地路径一致：

```
tool.sh accumulo_examples.jar examples.accumulo.SourceCountJob\
 -Dmapred.reduce.tasks=4\
 acled\
 /output/accumulo_source_count/\
 test\
 root\
 password\
 localhost:2181
hadoop fs -cat /output/accumulo_source_count/part* > source_count.txt
```

12. 保存并运行这个脚本，可以看到 MapReduce 作业在伪分布式集群上开始运行。

13. 当成功完成该作业，可以在你的工作目录得到文件 `source_count.txt`。输入命令 `cat` 可以看到每个消息源对应的统计总数。

工作原理

我们定义了类 `SourceCountJob`，实现了接口 `Tool`，通过类 `ToolRunner` 方便地远程提交作业。静态方法 `CachedConfiguration.getInstance()` 返回类路径中 Accumulo 正确的配置项给这里的 `Tool` 实例。

`run()` 方法使用 `AccumuloInputFormat` 解析连接 Accumulo 实例必需的参数。对于该作业，我们只关心从列族 `cf` 获得列 `src` 对应的每个键。默认情况下，扫描器只会返回包含列限定符 `src` 最近版本的键。如果希望计算出现在表中所有版本的键值对出现的消息源次数，需要在输入格式中配置 `maxVersions`。

随后，通过 `AccumuloInputFormat` 设置好作业，设置好用于计算每个消息源数目的 map/reduce 类。由于 reducer 类是简单地将整数进行相加，我们可以将其作为一个 combiner 的简单实现进行复用。删除已存在的输出文件。由于运行在伪分布式集群上，设置 reduce 任务的数目为 1。现在已准备好将作业提交至集群。

这里的业务逻辑和经典的例子 WordCount 十分类似。

类 `AccumuloInputFormat` 进行扫描并值返回满足列限定符 `scr` 的键值对。因此，任何进入类 `ACLEDSourceMapper` 的 `map()` 函数的键值实例，都需要聚集统计。我们可以简单的输出 1 表明在数据集中出现一次消息源值。

reduce 类 `ACLEDSourceReducer` 简单的标记了每个消息源的出现次数，并输出结果至 HDFS。

在 shell 脚本的最后，下载并合并不同 part 的文件至一个文件 `source_counts.txt`。现在，我们得到了包含换行符分割的消息源列表以及每个源的出现总次数的单个文件。